Special Publications
1

Bridge Construction Methods
Essentials and Critical Lessons

Practitioners' Perspectives

Editors
Pedro Pacheco
Filipe Magalhães
Gonzalo Osborne Gutiérrez

International Association for Bridge and Structural Engineering (IABSE)

ISBN: 978-3-85748-207-6 (print)
978-3-85748-208-3 (ePub)
978-3-85748-209-0 (eBook)

DOI: https://doi.org/10.2749/sp001

Publisher

IABSE
Jungholzstrasse 28
8050 Zürich
Switzerland

Phone: Int. +41-43-443 9765
E-mail: secretariat@iabse.org
Web: www.iabse.org

This book was produced in cooperation with Structurae, Dresdener Str. 110, Berlin, Germany (https://structurae.net).

Copyediting: Hilka Rogers
Layout & typesetting: Florian Hawemann

Note for the Readers

This book is designed to be a practical, "easy to read" resource to assist bridge engineers in various roles – including bridge owners, designers, constructors, or colleagues from academia. Focused on bridge construction, the book is driven by two simple and narrow questions: "How can I do it?" and "What is critical?". It is likely that many readers will not read the book in one sitting – but will refer to chapters as needed based on their specific needs and/or interests.

Regarding the question "How can I do it?" – in the first two chapters following the introduction, the reader will find a range of methods for the main structural elements of more common bridges and of special bridges. The methods are presented with varying levels of depth based on one simple criterion – if methods are assumed to be known by someone skilled in the field, the presentation is minimal; if not, the key characteristics of such methods are presented. However, readers should not expect a detailed description of each method – but they can certainly expect to explore a broad array of options and their fundamentals. This same principle is applied in subsequent chapters where classic aspects of construction are covered – risk, time, costs, logistics, and technology. In each case, only the fundamentals are presented unless, like in the case of "risks," the text includes content not found in the complementary bibliography. In brief, the answer to the presented question is conveyed simply: "What are my options; What are their fundamentals".

Regarding the question "What is critical?", this book includes a section of paragraphs with "lessons learnt and recommendations" at the end of every chapter. Again, the book follows a simple criterion, and – not adhering to any systematically casuistic approach or even a systematic analysis of the previously presented "fundamentals" – a set of empirically verified critical lessons are emphasised. This knowledge represents the practitioners' perspectives. The major advantage for readers is the ability to quickly and easily identify selective aspects – sometimes buried within tens or even hundreds of rules and/or specifications in the specialised bibliography – which should be highlighted due to their potential major implications or consequences. Some of these lessons may prevent significant deviations in costs, environmental impact, or time – i.e., if known and identified at the appropriate stage of a project from preliminary design to execution. More importantly - some of these lessons may save lives.

It is known that the Romans had the practice of recording incidents and accidents on wax plates and bringing them to a central building in Rome where these cases were discussed, and critical lessons were identified. This practice is likely one of the reasons Rome was one of the most significant "schools" of bridge engineering in history. Similarly, the purpose of this book is also to be a "wax plate" contribution to the bridge engineering community.

Pedro Pacheco

Acknowledgement

This book is the culmination of a multi-year, collective effort to forge a more holistic and informed perspective on bridge engineering, with a focus on construction. We set out with the primary objective to address the two pivotal questions, “How can I do it?” and “What is critical?”. The realization of this goal was made possible through the invaluable contributions of numerous experienced and distinguished bridge engineers.

Before proceeding, I must acknowledge the contributions of Jean-François Klein and Thomas Spuler, whose lives tragically ended too soon. Their significant impacts on bridge engineering worldwide continue and endure.

I extend my profound thanks to all the authors whose diligent and meticulous efforts helped answer the foundational questions across various domains of bridge engineering. There is significant complementarity among the authors: many are experienced engineers from the construction industry, others are prestigious designers, and still others are notable colleagues from academia. Special appreciation goes to those colleagues who continued to coordinate task groups even after retirement. **The authors are the ones “who did it”.**

My gratitude also extends to the members of Commission 4: Javier Muñoz Rojas, Filipe Magalhães, Gonzalo Osborne Gutiérrez, Akio Kasuga, António Sousa, Don Bergman, and Carlos Bajo, whose critical insights were instrumental in shaping the guidelines that steered this book’s development. Certainly, Filipe Magalhães and Gonzalo Osborne Gutiérrez, who are also editors of the book, deserve a distinguished note.

I am especially grateful to the Review Panel, adeptly led by the Chief Reviewer, Harshavardhan Subbarao. I owe particular thanks to Fabrizio Palmisano and Mohan Jatkar for their thorough reviews. Additionally, the insightful comments from Juan Sobrino, Andre Bance, Søren Lausten, and Craig Finley have enhanced this book.

Finally, on behalf of Commission 4 and the 17 task groups that brought this book to fruition, I extend heartfelt thanks to Tobia Zordan, head of the IABSE technical committee, for his invaluable advice and constant motivation. Likewise, I thank Tina Vejrum, President of IABSE, for her steadfast support and enduring commitment to excellence.

Pedro Pacheco

Preface

The function of a bridge is to provide a passage for traffic or other loads across some particular space like a natural obstacle or another man-made thoroughfare. The form of the bridge must express this basic function. Structural solutions for bridge designs, more than any other type of structure, grow out of the site. The best structural solution will be a design inherent in the constraints of the site itself which best fulfils the function of bridging the site. The task of the structural designer is to discover and accomplish that form in an economical and efficient way – to find a bridge design that can be easily constructed at a reasonable cost.

The way of building the bridge is decisively influential on a structural form not only for long-span arch or cable-supported structures but also for typical girder structures. It is important to find a way to minimise the impact of construction on the environment. That is why it is necessary for clients and designers to be familiar with all possible construction methods. They have to propose the structures, for which contractors can offer efficient construction.

While the designers need to have general information on construction methods, the contractors need to be profoundly knowledgeable about the technologies they use and aware of their advantages and disadvantages. This book gives us both, an overview of construction methods and thorough descriptions of individual technologies. And what is more, it gives us not only the required information but also 'critical lessons' gained during construction.

This book was written by experienced engineers who have made significant progress in developing new technologies, like for example, using prestressing in a new creative manner. Therefore, it is evident that they must be deeply knowledgeable about existing technologies as well as creative. And what is more, they want to share their gained experience.

That is why I strongly recommend this book to engineers who want to contribute to the development of our exciting profession.

Jiří Stráský

Contributing Authors

Written by IABSE Commission 4 members 2017–2024

Alp Caner
USA
WSP
alp.caner@wsp.com

Alp Caner is Vice President at WSP in New York.
He is involved in many bridge designs and bridge related research.

António Adão da Fonseca
Portugal
ADF Consultores
adaodafonseca@adfconsultores.com

Retired Professor of Bridge Design, Civil Engineering Department, University of Porto, and co-founder of ADAO DA FONSECA – Consultores de Engenharia, Portugal.

António Luíz Souza
Brazil/USA
Consultant
antonioluizsouza@gmail.com

Experienced Civil Engineer skilled in designing, planning, implementing and managing large construction infrastructure and building projects, working as an independent consultant based in FL, USA.

Arianna Minoretti
Norway
Statens vegvesen
arianna.minoretti@vegvesen.no

Senior Principal Engineer Technology and Development Department
Norwegian Public Roads Administration.

Artur Peixoto
Portugal
Apgeotecnia
artur.peixoto@apgeotecnia.pt

Civil Engineer, consultant in Geotechnical Engineering and Special Foundations.
Co-author of Deep Soil Mixing solutions applied to urban excavations. Co-founder of the company APGeotecnia.pt.

Carlos Bajo
Spain
Ferrovial
carlos.bajo@ferrovial.com

Head of Bridge & Civil Structures in Ferrovial construction Spain. MSc Civil Engineer and MASt Applied Economics. Specialist in bridge design and accelerated construction processes.

Cato Dørum
Norway
Statens vegvesen
cato.dorum@vegvesen.no

Working at the Technology and Development Department, Norwegian Public Roads Administration.

Filipe Magalhães
Portugal
FEUP
filipema@fe.up.pt

Associate Professor at the Faculty of Engineering of the University of Porto, Director of the Structural Testing Laboratory, with expertise on the testing and monitoring of structures.

Gonzalo Osborne Gutiérrez
USA
FLATIRON CORP
gosborne@flatironcorp.com

Bridge design and construction expert with 20+ years in heavy civil works, leading Flatiron's Bridge Technical Services in USA and Canada, with major experience across Europe and North America.

Hugo Coelho
Portugal
BERD
hugo.coelho@berd.eu

Civil Engineer, started practising in 2004, specializing in the areas of structural engineering, steel structures, bridge construction methods and risk assessment. Recognized as Expert in Structures by the Portuguese National Associations of Civil Engineering and member of the Spanish Association of Structure Engineers (CICCP).

Inês Ferraz
Portugal
BERD
ines.ferraz@berd.eu

Senior Structural Engineer and Product Development Manager at BERD, S.A. (Portugal) since 2005. Specialized on Modular Bridges and Bridge Construction Equipment.

Javier Martinez
Spain
ALE Heavy Lift
J.Martinez@ale-heavylift.com

Naval Architect, Polytechnic University of Madrid.
Founder of LASTRA, engineering & construction heavy lifting specialist in all civil & industrial sectors; later acquired by ALE and eventually by MAMMOET.
Retired in 2024.

Javier Muñoz Rojas Fernández
Spain
CFCSL
jmrojas@cfcsl.com

Javier Muñoz-Rojas Fernández, CEO at CFC (ARX Group), expert in special structures. Led 120+ projects including iconic bridges, stations, and complex infrastructure worldwide.

João Santos
Portugal
NADARA/LNEC
joao.santos@nadara.com

Head of Analytics, AI, Innovation & R&D at Nadara. Expert in data science, ML, sensing, monitoring, and leading multi-team innovation and R&D projects across domains.

John Kerkhoff
The Netherlands
Janson Bridging
john.kerkhoff@jansonbridging.com

Business Development and Sales Manager at Janson Bridging Netherlands BV.
BBE in Civil Engineering and working for Janson Bridging since 2002.

Johannes Veie
Norway
Statens vegvesen
Johannes.veie@vegvesen.no

Senior Principal Engineer Technology and Development Department;
Norwegian Public Roads Administration.

Jon Ruiz de Azúa
Spain
KSM Structural Monitoring
ruizdeazua.jon@gmail.com

Technical Director at Kinesia Structural Monitoring, Spain. Specialiced in dynamic and static monitoring of bridges during construction and service phases.

José Maria Sousa Soares
Portugal
BERD
jose.soares@berd.eu

Senior Structural Engineer and Project Manager in BERD.

Marcos Sánchez
Ireland
ARUP
marcos.sanchez@arup.com

Bridge Engineer. Director at Arup. Specialized in long span bridge design and construction.

Miguel Angel Astiz
Spain
CFCSL
maastiz@cfcsl.com

Consulting Bridge Engineer at CFC, Madrid.
Retired Professor in Bridge Engineering at Polytechnic University of Madrid.
Bridge designer specializing in wind and earthquake engineering.

Niculin Meng
Switzerland
Mageba
nmeng@mageba.ch

Structural Engineer, Head of EMEA and member of the Executive Committee at mageba Group.

Parthasarathi Jena
Portugal
Statens vegvesen
Parthasarati.jena@vegvesen.no

Senior Engineer
Complex Constructions Department;
Norwegian Public Roads Administration.

Paulo Lopes Pinto
Portugal
Univ. Coimbra
ppinto@dec.uc.pt

Assistant Professor in the Department of Civil Engineering at the University of Coimbra, Portugal. He is also a geotechnical consultant, specialising in foundation and seismic design.

Pedro Borges
Portugal
BERD
pedro.borges@berd.eu

Structural Engineer, specialized in Bridge Construction Methods and Bridge Construction Equipment, Director of Technical Solutions Department at BERD S.A.

Pedro Pacheco
Portugal
BERD
pedro.pacheco@berd.eu

Professor of Bridges, Civil Engineering Department, University of Porto, and founder of BERD, S.A.

Renato Bastos
Portugal
ADF Consultores
renato.bastos@adfconsultores.com

Structural Engineer, designer of bridges and buildings. Partner of ADÃO DA FONSECA – Engenheiros Consultores, Lda, Portugal.
Adjunct lecturer, Dept. of Civil Engineering, FCTUC, University of Coimbra, Portugal.

Simen Hellgren Holtberget
Norway
Statens vegvesen
simen.hellgren.holtberget@vegvesen.no

Senior Engineer
Technology and Development Department;
Norwegian Public Roads Administration.

Thierry Duclos
France
Vinci Construction
thierry.duclos@vinci-construction.com

Chief Structural Design Engineer. Retired Expert.

Torben Forsberg
Denmark
Forsbergs
tof@forsbergs.dk

Mr. T. Forsberg is a recognized structural engineer who have been leading the design of suspension bridges such as the cable structures of the Great belt Bridge (Denmark), the Hålogaland Bridge (Norway) and the Ozman Gazi Bridge (Turkey).

Xu Xiang
Norway
Statens vegvesen
xu.xiang@vegvesen.no

Senior Principal Engineer
Complex Constructions Department;
Norwegian Public Roads Administration.

Yasutsugu Yamasaki
Japan
DLSY Joint Venture
yasutsugu.yamasaki@dlsyjv.com

Mr. Y. Yamasaki is a World-renowned specialist engineer within construction of suspension bridges. Mr. Yamasaki has led the engineering tasks for many large suspension bridge constructions, such as Bosphorus II, Ozman Gazi Bridge and Canakalle Bridge in Turkey. Mr Yamasaki was also cable specialist engineer in charge of the Great Belt suspension bridge design in Denmark.

Table of Contents

https://doi.org/10.2749/sp001.ch01

1 Introduction

Pedro Pacheco[1] | Filipe Magalhães[1] | Gonzalo Osborne Gutiérrez[2]

[1] Portugal, [2] USA

1.1 Summary

This book is an overview of bridge construction considering different relevant perspectives and pointing out synthetic critical knowledge. The book was designed to be useful for anyone connected to bridge engineering – including bridge owners, designers, constructors, and also, academia. To achieve this goal, the book aims to answer 2 simple questions: '**How can I do it?**' and '**What is critical?**'.

Bridge engineering typically has numerous solutions worldwide showing a heavy influence of local context for selecting the most efficient means and methods. It is also influenced by the previous experience of contractors and designers in building similar bridges. Schedule constraints, required temporary works, preferred local materials depending on manufacture or logistics, or even qualified labour and equipment available locally will direct the way a structure is built. All these factors sum up the local best practice for each region, their influence being quite profound, in some cases. To some extent, bridge construction knowledge is organised in 'schools' (as tendencies). It is not rare that a bridge engineer simply does not know methods used in other regions of the world. The first question stated above is focused on a simple and concise contribution – presenting the readers with the essentials of different construction options gained from the authors' worldwide experience.

In bridge engineering, it is not infrequent that the best decisions are influenced by secondary aspects which are disengaged from pure design or pure construction matters. For example, the analysis of the logistics context, considered since the early stages of a project, may provide relevant benefits to the process for defining the ultimate solution. It is well known that this applies to risk, costs, and time. Additionally, it is quite common that the available bibliography and/or normative documents in each market are frequently quite enough to establish suitable construction procedures. However, before bridge engineers rise to a particular challenge, it is not easy for them to foresee it. As a matter of fact, facing a set of tens (or sometimes hundreds) of normative statements or recommendations, it may be difficult for the bridge engineer to know which ones will be critical for the specific structure under analysis. Moreover, there might be lessons learnt which have not yet been included in those documents. Hence, the second question stated above is focused on another simple and concise contribution – presenting the reader with a set of critical aspects in construction, which if known, may encapsulate significant benefits.

The book is divided into 5 chapters. Following the introduction, Chapters 2 and 3 are exclusively dedicated to bridge construction methods – the former reviews more general methods and the latter is devoted to special construction methods. Chapter 2 includes brief descriptions of construction methods for foundations, piers, bearings, joints and decks, and a section on modular bridges, which frequently refers to the same methods described in previous points. Chapter 3 provides an extremely condensed description of construction methods used in cable-stayed and extradosed bridges, arch bridges, suspension bridges, and floating bridges. In general, these 2 chapters (i.e., Chapters 2 and 3) do not reveal novelties and details are avoided – these chapters simply summarize in a compact text what can be found in fragments in a plethora of known books, papers, and normative documents, but are not easily found in a single simple document. As per the whole text, Chapters 2 and 3 end with a brief but particularly pertinent section on lessons learnt. Typically, 'lessons learnt' identify the main aspects of construction which may possibly seem obvious, but may have a disproportionately strong influence on construction success (or lack thereof). In some cases, 'lessons learnt' simply show a connection between the described methods with aspects of other perspectives of construction documented in the following chapters. In other cases, 'lessons learnt' reflect the merged experience of several authors with extensive experience in construction.

Chapter 4 covers three main topics for construction: risk, schedule, and cost. These items are closely related to the different construction experiences, as mentioned above. Particularly in this part of the book, a strong recommendation to the construction peer community is the use of numerical risk analysis tools in construction. In a nutshell, the general practices in the construction industry in numerous countries result in very lengthy risk analysis documents based on intuitive or empirical perceptions of risk. That is a serious risk in itself. It will be seen that certain studies sustain that this may be one of the most important causes of accidents in bridge construction – the lack of numerical risk analysis. A tested, simple, and practical numeric methodology is proposed in Section 4.1.

Chapter 5 presents an overview of the use of technologies in construction, from cast-in-place to precast solutions, including general equipment and heavy-lifting methods. Monitoring systems in specific construction solutions complete the content of this chapter.

In essence, this book distils a holistic view of the essentials of bridge construction, spotlighting critical facets reported by a blend of diverse construction experiences from around the globe.

1.2 Summarised State of the Art and Applicability of Construction Methods

The present state of the art in terms of applicable means and methods across different span length ranges can be summed up in Fig. 1.1 but is continuously under evolution (adapted from [1]). In Fig. 1.1, the most frequent span length ranges for each construction method are shaded in green (with yellow and orange being less common). The vertical lines subdivide the span ranges more common in multi-span viaducts (lighter grey or white) from the ones that are less common (darker grey) in such typology of bridges/viaducts.

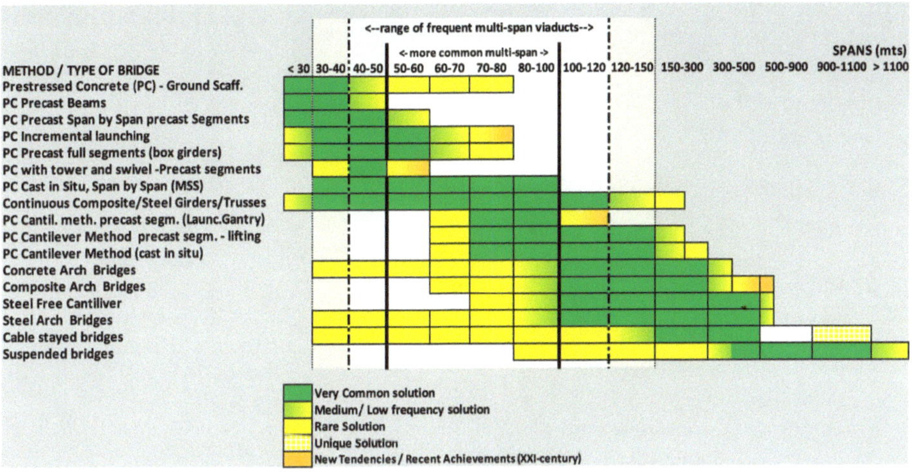

Fig. 1.1 Summary of Construction Methods – versus applicable span ranges [1]

1.3 Essentials on Decisions about Construction Solutions

For some bridges, it is well known that construction stages are potentially more critical than the service stages and that the construction solution has relevant impacts on costs and time [2].

The preliminary directions for the construction solution are sometimes provided by the bridge owner and are systematically and primarily established during the design stage. However, there are still some options that can be defined and refined during construction.

To ensure a reliable and informed decision-making process it is crucial for the decision-maker to have a 360-degree vision of the process, considering 1) the available construction methods; 2) interface knowledge areas in construction, namely risk, cost, and time; and 3) the available technology.

It is also relevant for the decision-maker to have a clear and well-defined vision of a) what the key goals are; 2) what the key decisions to be made are; and 3) what the main variables of the context are.

1.3.1 Key Goals in a Construction Solution

The Key Goals of a Construction Solution require an appropriate evaluation of the following topics:

- Safety and risk
- Quality
- Time
- Economics
- Sustainability and ecological footprint (in particular, CO_2)
- Lifecycle and maintenance
- Aesthetics (through design)
- Historical, cultural, and landscape integration (through design)
- Other environmental impact.

These key goals may certainly have different weights in the decision-making process for each particular project and that variation may obviously lead to the selection of a different procedure if different importance factors are applied to the various key goals.

1.3.2 Key Decisions for a Construction Solution

The key decisions of the construction solution are as follows:

- What constructive solution (design stage)
- What logistic concept
- What site facilities plan concept
- What construction methods and means (following the design solution)
- When or what planning approach (within bridge owner deadlines)
- What procurement and site supply chain management approach
- What skills of the core team
- What potential impacts on the environment.

The decision-making process will be strengthened if these questions are accompanied by the **2 nuclear questions** of this book:

How can I do it or What are the existing solutions?

What is Critical or What lessons have been learnt?

1.3.3 The Context of Technical Issues, Market and Logistics

The context characterisation – technical issues, logistics, and local market – is fundamental to pursuing an appropriate and rational decision-making process. Supplementing the bridge programme, the key context general factors are as follows:

- Local physical conditions (atmospheric, seismic, orography, etc.)
- Market restrictions and opportunities
- Local social and political conditions
- Local know-how and import opportunities
- Accessibilities of the bridge location
- Related logistical context (possibly global)
- If required, international trade conditions and restrictions.

It is thus fundamental to perform a thorough and comprehensive characterisation of the context before starting the decision-making process. To an experienced bridge engineer, when orographic conditions are very similar in 2 places, even if other technical conditions are similar, this does not imply that the best solution is the same. That is one of the reasons why in several languages, bridges are referred to as 'art construction'.

1.3.4 Remarks on 'Lessons Learnt'

Throughout the book, the reader will find practical remarks presented as critical lessons and recommendations at the end of each chapter. Summarising the each chapter, there are a few bullet points that, sometimes being muti-disciplinary issues, may be crucial to address when defining the construction solution.

- Have a good knowledge and understanding of the different construction methods and their merits and challenges – as presented in Chapters 1 and 2
- Study and verify that each construction method has its own merits and disadvantages – it is important to evaluate the best solution in each case for the context and key goals
- Determine if the core team skills fit the requirements for the specific bridge to be built

- Develop a quantitative numerical construction risk analysis (a qualitative analysis may be insufficient and risky in some cases)
- Acknowledge that for large span bridges, i.e., cable-stayed and extradosed, suspension and arch bridges, there is no way to find an appropriate solution if a deep understanding of the construction method is not gained since the early stages of preliminary design – much before the construction stage starts
- Acknowledge that monitoring the structure during construction provides opportunities to improve solutions and may prevent surprises, increasing quality and safety
- Stay informed of the existing equipment in the market and carefully evaluate the corresponding technological limits
- Perform integrated bridge-equipment interaction analysis, when applicable
- Develop sustainability practices, for example, an early-stage CO_2 emissions evaluation
- Perform laboratory or full-scale tests when using innovative solutions
- Acknowledge the logistic context and the research on the forecasts for the logistics context when defining the construction solution and supply chain strategy.

Some of these remarks, if not addressed, can jeopardize the selected solution. For example, both quantitative risk analysis and sustainability requirements (such as CO_2 emissions control) during construction might be legally enforced in various countries in a matter of a few years (and in some, they already are).

References

[1] Pacheco et al (2015) 2015, "Multi-Span Large Decks – the Organic Prestressing Impact", Multi Span Large Bridges Conference, pp 103–124, Porto, Portugal, CRC Press – Taylor & Francis Group

[2] Wai-Fah Chen, Lia Duan, "Bridge Engineering – Construction and Maintenance", Engineering Series, 2000, CRC Press

2

General Bridge Construction Methods

2.1 Essentials on Foundation Construction

Paulo Lopes Pinto | Artur Peixoto

Portugal

2.1.1 Introduction

The construction of bridge foundations is a challenging and complex activity that requires specialised designers and contractors. This section presents a brief overview of some of the relevant aspects of foundation construction. The bibliography offers a list, necessarily incomplete, of some of the key norms and texts where the reader may find useful insights [1–15].

The ground investigation (GI) is a key stage for bridge foundation design and for the selection of appropriate construction procedures. The objective of a GI is to produce a ground model where the relevant geological characteristics are represented, such as the material sequence, structure, variability, and hydrogeological characteristics, among others. This should be preceded by a desk study followed by a site reconnaissance. Both facilitate the assessment of a first model and the definition of the GI plan. This plan may include several stages of in-situ and laboratory testing, as the gathered data is processed and interpreted. Typically, the preliminary bridge design is developed in parallel with this stage and may also dictate the need for further testing to determine the ground parameters required for the foundation design. The results of this process are presented in the GI report and are used to develop the geotechnical design model (GDM) where the representative (characteristic) values of the ground properties are displayed.

The GI includes the execution of boreholes with samplers driven into the ground or holes drilled with rotary tools such as augers or core drills. The materials found are identified and samples are collected for laboratory testing. Standard penetration tests (SPT) are performed routinely in the boreholes to characterise soils, but other testing techniques are available and recommended as the quality of the acquired data is higher and the correlations to determined ground parameters are more reliable. Cone penetrometer (CPT), pressuremeter (PMT), and dilatometer (DMT) tests are some of the most used in-situ tests available.

2.1.2 Spread Foundations

Spread foundations are the most economical solution for bridge foundations if they are suitable for the ground and load conditions. This choice requires a proper GI to assess the bearing capacity and predict the settlements of the foundation. Settlement control is particularly relevant to prevent functional and structural anomalies due to differential settlements between adjacent footings. Ground improvement may be an option when the existing ground characteristics are not suitable for the use of spread foundations. Ground improvement techniques aim to increase shear strength and modulus, reduce permeability, and minimise liquefaction potential. These include densification, consolidation, and reinforcement methods, such as dynamic compaction, vibro-compaction of cohesionless soils, installation of wick drains, and surcharge to pre-consolidate clayey materials.

Regarding construction issues, four aspects need to be considered: i) excavation methods and equipment to reach the desired foundation level; ii) viability of excavation slopes versus the need to shore the ground as the excavation proceeds; iii) existence and impact of groundwater on the excavation process and stability; and iv) the existence of obstacles to the excavation. These issues need to be addressed in the GI report so that both the designer and the contractor are familiar with them before beginning the work.

In favourable soil conditions, mechanical excavators are used with good efficiency, but when the ground is more rigid, hydraulic hammers are needed and the excavation time and cost increase significantly. In rock masses, excavation may require the use of explosives. In this situation, the rock blasting plan must be carefully designed to minimise ground disturbance. The selection of the excavation method may be based on p-wave velocity, rock unconfined compressive strength (UCS), spacing of the discontinuities (joints, bedding, faults), and rock quality designation (RQD). P-wave velocity is determined through seismic refraction tests and equipment manufacturers provide rippability charts based on p-wave velocity for different types of rock. Franklin (1974) proposed a chart relating unconfined compressive strength and discontinuity spacing to assess rock mass excavatability.

The inclination of the excavation slopes should be defined based on the shear strength parameters of the materials intercepted which are presented in the ground investigation report. After the footing is concreted and the concrete has reached the strength specified in the design, the fill should be carefully compacted. When passive earth pressure is to be mobilised to prevent sliding, the compaction procedure and control should guarantee the required resistance.

If temporary shoring is required, it should be designed to resist the earth pressure acting on the external face. Zones of loose or weak soil and the existence of high groundwater should be identified in the ground investigation report and accounted for in the geotechnical design of the shoring. It should also be noted that if timber shoring is used and not removed after concreting the footing, it will decompose over time and larger horizontal displacements will be required to mobilise the passive resistance.

When excavating below the water table two alternative methods are to be considered: lowering the groundwater level by pumping the water from the exterior around the excavation zone or building a cofferdam (see Fig. 2.1) and pumping the water from the interior as the excavation proceeds. The choice requires knowledge of the hydraulic conductivity of the ground materials. This can be determined by in-situ testing (LeFranc, Lugeon, among others). Problems such as hydraulic instability of the bottom of the excavation and soil softening must be prevented.

Fig. 2.1 Cofferdam (left – external perspective, right – internal perspective) (photo: Mota-Engil)

Obstacles such as utilities, boulders, or old constructions may be encountered during excavation and need to be identified prior to construction. Provisions need to be incorporated in the design to prevent delays and litigation.

The size of many bridge footings means that considerable volumes of concrete must be placed. This may cause high temperatures during curing, which can lead to fissuring. In these cases, it is important to control the concrete temperature.

2.1.3 Cofferdams

By definition, a cofferdam is a temporary structure designed to retain water and/or soil during excavation for the construction of various types of structures buried below the water table, enabling work in a dry and safe environment.

Cofferdams are ancient works that have evolved due to the challenges faced in construction processes. These emerged in the speciality of hydraulic works, to meet the need to perform tasks in sensitive areas, such as in rivers, lagoons, and coastal zones, as well as deep excavations below the water table in sandy soils.

Cofferdams can be made from various types of materials, such as earth, rock, wood, steel, and concrete. The most common are usually constructed under the shelter of steel sheet piles, braced by steel walling and props, generally dismantled after the completion of the work.

The basic requirements of any cofferdam are to be watertight and stable against hydrostatic pressure and soil pressures, and to allow pumping its interior, whilst maintaining the stability of the bottom of the excavation. Most frequently cofferdams require the installation of a hydraulic pump.

Fig. 2.2
Types of Caissons

a. Box Caisson (courtesy of ETERMAR)

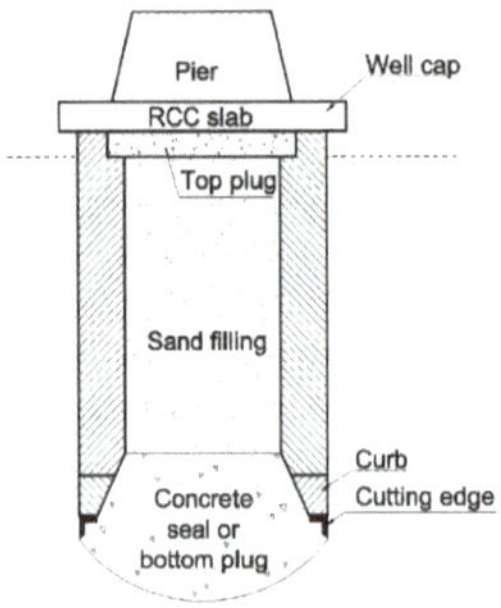

b. Open caisson or Well

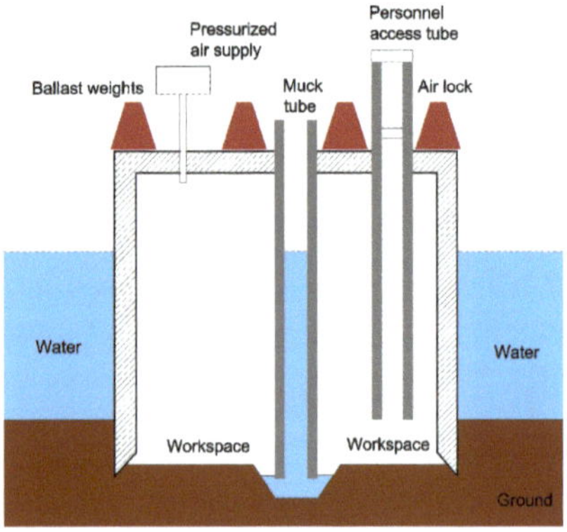

b. Pressurized caisson

2.1.4 Caisson or Well Foundations

Caissons are hollow constructions comprising one or more cells and are sunk into the ground or water to form a permanent shell of a deep foundation. They provide a very rigid foundation, with good lateral resistance and bearing capacity and are particularly suited to withstand large lateral loads and moments. If placed at a suitable depth with proper protection, caissons can withstand the effects of scouring due to water flow.

Caissons are made of steel or concrete and can be built onshore, floated to the location of the bridge pier, and carefully sunk onto the sea or river bed. Caissons that have a slab at the bottom, with the top open, are referred to as box caissons (Fig. 2.2a). Once placed in the final position they are filled with sand, gravel, or concrete. Caissons with an open bottom (Fig. 2.2b) are provided with a cutting edge and are used if there is the need to excavate through the ground to reach the bearing layer, typically the bedrock. This type of foundation in which construction and sinking proceed

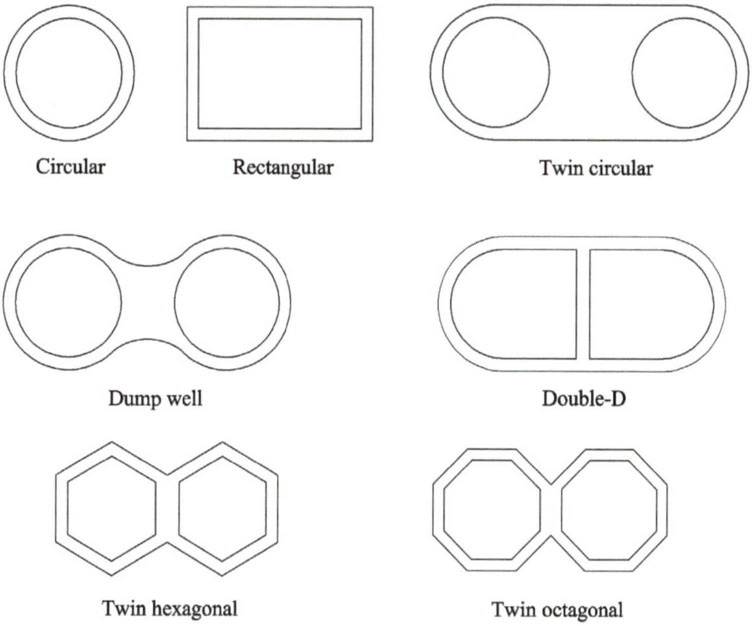

Fig. 2.3 Shapes of Well Caissons

simultaneously is known as a well foundation. Air-pressurised caissons (Fig. 2.2c) were used in the construction of several bridge piers, with high costs in terms of the life and health of the workers. With the advances of construction equipment, this practice is currently seldom used.

The shape of well caissons depends on the geometry of the piers and abutments, forces acting on the base of the pier, water flow, cost of sinking, and the possibility of misalignment during the descent of the well. As shown in Fig. 2.3, typical shapes include circular, rectangular, twin circular, dump well, double-D, twin-hexagonal and twin-octagonal wells.

The construction method for box caissons is as follows:

- The ground surface is levelled at the depth where the caisson is to be supported. If the caisson is to be supported by a group of piles, the piles are installed with the top at the same level and a pile cap is constructed. The caisson is usually supported by a base of sand
- The caisson is prefabricated onshore and floated to the pier location, where it is sunk with its position carefully controlled
- The cells are filled with sand, gravel, or concrete
- The top of the caisson is sealed with concrete and a cap of reinforced concrete is constructed, serving as the foundation for the pier.

Box caissons are a good solution when a resistant layer is found at shallow depth, the ground surface is already levelled or is easy to dredge, the water flow is slow, and the water is not very deep.

Open caissons (wells) on the ground or in shallow water (up to about 5 metres) may be constructed using the following method:

- An initial pit is dug on the ground surface, with the shape and size of the well, 3 to 4 metres deep. If the water table is above the excavation, a cofferdam may be installed, and the water is pumped from its interior to allow for a dry excavation
- The cutting edge, made of steel or concrete, is placed at the bottom of the pit.
- For reinforced concrete caissons, steel reinforcement and shuttering are placed above the cutting edge. Concrete is poured in small lifts and once it has gained enough strength, the caisson is sunk
- The sinking of the caisson occurs due to its own weight as the soil from inside the caisson is excavated.
- The next level of the caisson is concreted
- As excavation progresses, the caisson sinks, overcoming the side resistance between the caisson wall and soil. If the side resistance is higher than the caisson's weight, making this sinking difficult, additional weight may be placed at the top of the caisson
- Once the caisson reaches the final depth, a thick concrete plug is placed at its bottom, using a tremie pipe
- After the concrete plug has hardened, water is pumped out from the well, which is then filled with sand, gravel, or concrete
- A concrete plug is poured at the top of the sand fill and a reinforced concrete cap is constructed. This cap is the foundation of the pier.

Advantages of open caisson foundations include:

- The ground is inspected as the excavation progresses, together with boreholes that may be drilled at the bottom, which allows the decision regarding the final depth to be made during the process
- Obstacles such as boulders may be overcome during excavation
- The sinking of the well does not cause vibrations, limiting damage to the adjacent structures.

When there are concerns about the stability of the bottom of the excavation, provisional piles may be installed to provide additional resistance.

For deep waters, the ballast added to the self-weight might not be enough to sink the caisson, and the construction and installation of these caissons vary on a case-by-case basis. Alternative techniques are the jet air method, the jet water method, the jack-down technique, the vertical cutting edge method, or the pneumatic caisson technique which could can be used in waters up to 40 meters.in depth. All these techniques help the sinking operations by dredging or pushing down the caisson with the help of jacks, always in a controlled environment to ensure the stability and alignment of the caisson.

2.1.5 Deep Foundations

2.1.5.1 Introduction

When it is not feasible to use spread foundations, as is the case for many bridge foundations, piles, drilled shafts, or barrettes must be installed to provide appropriate capacity and to control settlements and horizontal displacements. Piles may be driven or bored, and often a combination of both techniques is used. This report only refers to the most used pile types: driven and bored piles.

When the ground is made of loose, saturated, or cohesionless materials, it may be susceptible to liquefaction under seismic loading. In such cases, soil improvement techniques may be required to densify the ground and mitigate the risk of damage due to loss of strength caused by liquefaction. One of these techniques is the installation of stone columns in the vicinity of the bridge pier. Vibroflotation is also used for the same purpose. The risk of scour should also be assessed and provisions to mitigate its effects include placement of rock blocks on the river bed, close to the foundation, or the assumption of a 'sacrificial' pile length that is not considered for bearing capacity calculations.

Piles installed in clayey soil undergoing consolidation settlement may suffer down-drag forces caused by negative skin friction as the soil settles more than the piles. This issue requires proper attention both in pile design and construction measures to mitigate the risk of excessive pile settlement and axial stress.

2.1.5.2 Main Types of Piles

2.1.5.2.1 Driven Piles

Driven piles, also defined as displacement piles, are deep foundation elements installed using impact or vibration hammers to a design depth or resistance. Depending on their distance from the shore, they could be considered as land piles or marine piles.

Driven piles are typically circular or square precast concrete piles (Fig. 2.4) or steel piles with tubular or H sections. They are defined as friction piles or end-bearing piles with respect to load transmission and functional behaviour. Fig. 2.5 depicts pipe piles driven with a vibratory hammer and with the interior excavated with a grab.

Fig. 2.4 Group of driven precast concrete piles

Fig. 2.5 Pipe pile driven with a vibratory hammer and excavated with a grab

Common uses of driven piles

- They can be used for all types of construction, particularly in aggressive soil conditions
- They are well-suited to sites where the ground conditions are highly variable as they are driven to a set or pre-determined resistance
- They are a good foundation choice when very thick layers of soft soil and/or a high water table are encountered which could be problematic for a traditional drilled pile

- They are commonly used to work in remote areas
- Small pile sizes are typically used with low pile demands (abutments, walls, or short bridges). These could also be used in difficult access areas like steep slopes where medium-sized pile equipment has limited access
- Medium-sized piles are commonly used for all types of bridge and wall foundations with medium demands
- Large-sized piles are mostly used for pylons, river bridge foundations with scour issues, or in high seismicity areas and liquefiable soils. These piles have a high capacity for axial loads working as end-bearing piles and for lateral loads. They could be designed as monopile pier foundations (for example combined with hammer piers) in areas with reduced space or right-of-way which prohibits standard pile cap solutions. Medium and large-sized piles typically require fully or partially excavating and filling the pile with reinforced concrete to increase structural capacity.

Process

Piles are driven with modern, free-fall equipment, using a hammer raised either by a simple cable system or the most advanced high-performance hydraulic driving methods. This equipment is completely autonomous (requiring no auxiliary components) and mounted on crawler cranes for easy movement.

Driven piles gain geotechnical capacity efficiently by displacing the soil around the shaft and compacting the soil at the toe during installation. The energy to drive the piles is provided by either a high-frequency oscillating hammer or a percussion hammer. If penetration of dense soil is necessary it is often desirable to reduce the required installation energy, so pre-drilling may be required for the pile to reach the design depth.

The finished foundation element is designed to resist the imposed compressive, uplift, and lateral loads. Driven piles can also be used to provide lateral support for earth retention walls.

Advantages of driven piles

- Cost-effective and efficient design solution
- Displacement pile resulting in zero spoil, making it an ideal solution for contaminated sites
- Fast and clean operation
- Low bearing pressure for platform design
- Easily fabricated for high structural strength
- Suitable for most ground conditions
- High degree of confidence around pile capacity and integrity in real-time through installation monitoring and dynamic testing
- Range of noise mitigation measures available for sensitive areas
- Self-erecting rigs for fast and efficient mobilisation.

Quality assurance

Piles should be driven to the required ultimate pile capacity or the required ultimate pile capacity and minimum tip elevation if specified, or the specified tip elevation.

The ultimate pile capacity is usually confirmed by achieving the specified blow count which is the number of hammer blows required for a specified penetration.

a. Drilling equipment with telescopic Kelly bar

b. Temporary casing

c. Rebar cage placement

d. Concrete pouring

e. Piles tops ready to be demolished

f. Pile tops after demolishing

Fig. 2.6 Construction process of bored piles

Wave equation analysis is sometimes used to establish a driving criterion in preparation for performing a static or dynamic test. The wave equation is also useful to evaluate whether the maximum driving stresses caused by the proposed hammer are acceptable.

Determination of ultimate pile capacity

The required ultimate pile capacity is a function of the specified load evaluation method (calculation, static testing, dynamic testing, or rapid load testing).

2.1.5.2.2 Bored Piles (Drilled Shafts)

Bored piles require the excavation of stable holes where the reinforcement cage is placed and concrete is poured (in the case of continuous flight auger (CFA) piles, concrete is poured before the installation of the reinforcement). The construction process for bored piles with temporary casing is shown in Fig. 2.6. Casing (permanent or temporary) or dense fluids, such as bentonite or polymers, are used to stabilise the ground during excavation. The excavation and concreting phases are critical to obtaining sound and competent

a. Soil auger

b. Drilling bucket

c. Rock coring equipment

Fig. 2.7 Drilling equipment for soil and rock

bored piles. Steel pipes used as casing may be installed only for the top section where the ground is less stable, and the lower section is dug without casing if the ground is cohesive and does not cave. Particular care is required if artesian water pressure is found onsite. When excavating in such ground conditions, the water may burst in as the hole nears the layer where the water pressure is higher than the hydrostatic pressure, and consequently, the base and shaft resistances will be compromised. When this problem is foreseen, the fluid pressure in the bottom of the hole during excavation and concreting should be kept higher than the artesian pressure. Concreting of the bored piles must be performed ensuring that no concrete segregation occurs and preventing the soil from falling into the concrete. Spacers are placed in the reinforcement cage to maintain concrete cover within the specified limits. When the concrete has reached the strength specified in the design, the pile top is demolished and the rebars are connected to the pile cap.

Bored piles typically have diameters ranging from 400 to 2500 mm and may support very high loads. They are recommended when layers with cobles or boulders must be crossed, and when rock sockets are required. There are several drilling tools, such as shown in Fig. 2.7 for soil and rock, that make the use of bored piles possible for a very wide range of ground conditions. If the soil is hard, reverse circulation drilling rigs (RCD) can be applied.

There is a particular type of drilled piles which have an enlarged base (Fig. 2.8), also known as '*Tubulão*' in South America, particularly in Brazil. This type of pile construction is distinguished by its enhanced resistance, a significant improvement over traditional piles with a constant diameter. The construction process begins with the excavation, followed by the assembly of steel reinforcement, tasks typically performed manually by labourers. Bracings or compressed air are employed to ensure the stability of the hole during this phase. However, this method entails significant risks, as soil may cave into the hole and bury the labourers. Consequently, it is imperative that such construction is only undertaken following a thorough and careful risk analysis. The implementation of exceptionally stringent safety procedures is crucial to mitigate these risks, emphasising the importance of prioritising safety in these construction projects.

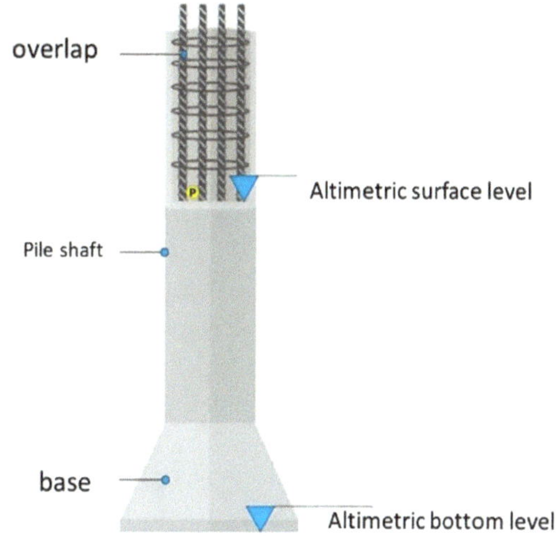

Fig. 2.8 Drilled Pile with enlarged base (photo: www.pedreirao.com.br)

2.1.5.3 Integrity Testing

At the end of the pile installation, it is necessary to check that its integrity is not compromised. Ensuing complications of driven piles include pile breaking and concrete being crushed or cracked during driving on hard or soft ground. Bored piles may suffer from concrete segregation, soil mixing with concrete, and soil caving into the borehole. These problems will compromise pile behaviour and capacity. Therefore, they need to be identified at the construction stage when corrective measures are still possible.

Pile integrity may be assessed through non-destructive testing, where sonic echo and cross-hole testing play a major role. Other techniques such as thermal profiling and gamma-gamma logging are also available but are used less often than the previous two.

Sonic echo testing uses the impact of a hand hammer at the pile top to generate a stress wave that propagates through the pile, is reflected at the pile tip, and returns to the top. An accelerometer is placed at the pile top and records particle acceleration. Compressional wave velocity depends on the pile material (in concrete, compressional wave velocity ranges from 3000 to 4000 metres/second and in steel it is equal to 5100 metres/second) and is used to determine the time required for the stress wave to travel a distance equal to twice the pile length. If an early arrival of the reflected wave is detected, this means that there is an anomaly. On a driven pile, it is a sign of a cracked pile and on a bored pile, it indicates a deviation from the intended pile cross-section or poor-quality concrete. This type of test has limitations when the piles are long or have a large diameter because the impact energy is not enough to obtain a clear toe reflection. An alternative procedure to solve this problem is the use of cross-hole testing, where several parallel pipes are placed next to the longitudinal steel reinforcement before concreting with known distances. During the test, a wave generator and a receiver (geophone) are lowered and raised in two pipes, and the time required for the wave to travel from one pipe to the other is determined. When this time is deviating from the expected value, it is a sign of poor-quality concrete. The process is repeated for the remaining pairs of pipes to assess the integrity of the entire pile.

Thermal integrity profiling (TIP) is a method based on the fact that concrete curing is an exothermic chemical reaction, where the temperature rises during the initial phase of the process and then gradually reduces as the heat is transferred to the surrounding environment (i.e., air at the pile top, and the ground on the sides and tip). On a sound bored pile, heat transfer patterns can be predicted and compared to the measured temperature. Thermal cables, capable of measuring temperature, are attached to longitudinal bars of the reinforcement. They are used to monitor the temperature distribution and variation over time, along the drilled shaft.

Defective concrete results in a localised decrease in temperature which can be identified by the measurements along the pile length. The method can be used to verify pile diameter, concrete cover, and reinforcement cage alignment. The main advantage of thermal profiling is that it permits the assessment of concrete quality in the entire cross-section of the bored pile for a short period (2 to 3 days) after pile installation. The downside is that it is not possible to perform this test after the temperature has dropped. This limits its use to the early stages of bored pile construction. Thermal integrity profiling cannot be used to assess the effectiveness of the remedial works that may have been performed. The other limitation consists of the difficulty of assessing the concrete quality near the pile tip, where concrete mixing with soil or drilling fluids or poor cleaning of the bottom of the borehole may impair the base resistance. Other non-destructive integrity tests may be used together with thermal integrity profiling to overcome these limitations.

2.1.5.4 Load Testing

Load tests are essential to evaluate pile behaviour under axial and lateral loads and to confirm the design and construction methods for the bridge foundation. If performed before the installation of production piles, load tests can be used to improve the design and reliability of the foundations, resulting in lower factors of safety (or larger resistance factors for the load and resistance factor design or LRFD approach). It should be noted that provisions on most codes (AASHTO, Eurocode 7, among others) allow for the reduction of the safety factor when load tests are performed.

When side and base resistances are measured separately it is possible to determine the resistance of piles with different cross-sections installed with the same procedure as the test pile. Tests may be carried to the failure load or to a proof load specified to verify the resistance and/or serviceability requirements of the design. Piles tested for soil failure provide valuable additional information.

Static Load Tests

Static load tests are the most reliable method to measure axial load-transfer performance of piles. As a result, piles that are tested under this condition require a lower factor of safety (or larger resistance factor), leading to potentially significant cost savings. The load can be applied at the top of the pile, in compression or tension, or at the bottom through a preinstalled hydraulic jack placed at the pile tip (Bi-Directional or Osterberg tests). Top-down load tests require a reaction system (kentledge, beam supported on piles, ground anchors), hydraulic jacks and instrumentation (load cell, dial gauges). Internal instrumentation (tell-tales, strain gages) is required to measure side and base resistances separately.

For driven piles, the pile's ultimate pile capacity may increase (soil setup) or decrease (relaxation) after the end of driving. Therefore, it is essential that static load testing be performed after equilibrium conditions in the soil are re-established.

High Strain Dynamic Load Tests

Dynamic load tests use the force applied during the blows of the pile hammer on a pile instrumented near the top with a pair of accelerometers and another pair of strain transducers.

Ultimate pile capacity at the time of testing is determined using a signal-matching analysis of the data. This technique can also be used to evaluate the reliability of wave equation analyses for driveability by measuring pile stresses during driving and the performance of the hammer in transferring energy to the pile.

Rapid Load Tests

Rapid load tests are performed by applying a fast load pulse for a longer duration than in dynamic testing. The load is applied by dropping a large mass on a spring cushion over the pile top or by placing a heavy mass on top of the pile and accelerating both pile and mass using combustion gas pressure. The method does not require a reaction system and is able to apply a high load at the pile top. The rate effects must be considered in the determination of the pile capacity.

A limitation of both high-strain dynamic and rapid load testing is that they are unable to assess the creep behaviour of piles installed in clayey soil.

2.1.6 General Recommendations

This chapter provides a brief summary of some of the relevant aspects that need to be addressed in the challenging world of bridge foundation construction. The design and construction of bridge foundations require a ground characterisation to be able to provide the parties involved (owner, designer, contractor, financier, etc.) with a ground design model that is as comprehensive as possible. This ground model will be used to define the type of foundation that is considered most appropriate for the bridge and for the site where it is to be built. In order to achieve this goal, a thorough geological/geotechnical ground investigation must be designed, performed, and interpreted by qualified experts. The decision regarding the type of foundation, its detailed design, and the definition of construction procedure and quality control must be made by experienced engineers, supported by contractors with relevant experience and know-how.

2.1.7 Lessons Learnt and Recommendations

Always taking into account the previous point of the General Recommendations, the following points may be very useful to be addressed in bridge construction practices:

- It is recommended to always have a Plan B for a singular pile failure in a pile block.
- Accuracy is imperative in foundations with precast pile blocks
- Requirement of a deeper investigation in clay soils with hard layers (N-values of Standard Penetration Testing or N_{SPT} >60) – as in some

cases soft layers may appear bellow hard soil occurrences

- Avoid cold joints in large pouring operations (spread or well foundations and pile blocks)
- For bridges where the dimension permits such options, it is recommended that the constructor erects a few test piles for better control of the construction process, which is even more important in the case of demanding geometric requirements
- For bridges where the dimension permits such options, it is recommended that the constructor implements load tests to evaluate the possibility of optimising the piles. This option must be approved by the bridge owner and/or designer
- Sometimes huge or relevant economic savings and a CO_2 footprint emission reduction may be allowed by implementing load tests on piles and reviewing the calculation assumptions, if possible and applicable.

References

[1] Wai-Fah Chen, Lian Duan, Bridge Engineering Handbook – Substructure Design, CRC Press, 2014

[2] Petros P. Xanthakos, Bridge Substructure and Foundation Design, Prentice Hall, 1995.

[3] V. Modeer, R.K. Bharil, Bridge Foundations, Innovative Bridge Design Handbook, Butterworth-Heinemann, pp. 729–750, 2022.

[4] Matos Fernandes, M., Analysis and Design of Geotechnical Structures, CRC Press, 2020.

[5] AASHTO LRFD Bridge Construction Specifications, 4th Edition, with 2020 Interim Revisions.

[6] AASHTO LRFD Bridge Design Specifications, 8th Edition, 2017.

[7] Brown, D.A., Dapp, S.D., Thompson, W.R. and Lazarte, C.A. CEG 8 – Design and Construction of Continuous Flight Auger (CFA) Piles. FHWA-HIF-07-039, 2007.

[8] Brown, D.A., Turner, J.P. and Castelli, R.J. GEC 10 – Drilled Shafts: Construction Procedures and LRFD Design Methods. FHWA-NHI-10-016, 2010.

[9] Day, R.W. Foundation Engineering Handbook. Design and Construction with the 2009 International Building Code. McGraw Hill, 2010.

[10] EN 12699:2015 – Execution of Special Geotechnical Works – Displacement Piles, CEN.

[11] EN 14199:2015 – Execution of Special Geotechnical Works – Micropiles, CEN.

[12] EN 1536:2010 – Execution of Special Geotechnical Works – Bored Piles, CEN.

[13] Franklin, J.A. (1974). Rock quality in relation of the quarrying and performance. 2nd IAEG, São Paulo, Brazil.

[14] Hannigan, P.J., Rauche, F., Likins, G.E. Robinson, B.R. and Becker, M.L. GEC 12 – Design and Construction of Driven Pile Foundations, vol. I, II and III. FHWA-NHI-16-009, 010 and 011, 2016.

[15] Kimmerling, K. GEC 6 – Shallow foundations – FHWA-SA-02-054, 2002.

2.2 Essentials of Pier Construction

Renato Bastos
Portugal

2.2.1 Introduction

The section presents the essentials of pier construction. Table 2.1 presents an illustrated list of the most common erection methods for piers. The text is then essentially focused on concrete columns built on site using climbing or sliding formwork or on piers assembled with precast elements with a short reference to other types of piers.

In any construction method, the construction time cycle is heavily influenced by access, pier shapes, and other conditional factors. In the crane climbing method, the planning (further detailed in Section 2.2.2) may allow for a cycle of 2 to 4 days to be commonly achieved for each lift.

During the construction of bridge piers, in addition to the conventional formwork system used for elements of lesser height, climbing and sliding formwork systems are typically employed for columns exceeding 15 metres in height. The most commonly used methods are climbing formwork—both self-climbing and crane-assisted climbing—and sliding formwork.

2.2.2 Climbing Formwork

Climbing formwork is a specialised type of formwork designed for vertical concrete structures that progresses with the construction. This method is predominantly used in the construction of bridge piers, towers, skyscrapers, and other tall vertical structures. It permits the repeated use of the formwork for successive identical sections of the structure as it grows. Typically, this type of formwork includes work platforms and scaffolds for construction crews and may also provide areas for machinery as well as screens for weather protection. It can be fully enclosed while remaining modular to adapt to the evolving vertical structure.

There is a vast array of climbing formwork systems available, each with its specifications and benefits. Owing to this variety, a thorough analysis is crucial to choosing the appropriate system for a specific project [1].

- Crane climbing formwork: This system is usually composed of two panels (one on each side of the wall being concreted), moving in tandem and connected to a rigid steel structure with fastening elements. The formwork is raised by one or more cranes once the concrete has sufficiently hardened.
- Self-climbing formwork: Similar to crane climbing formwork, this system includes a hydraulic mechanism that enables it to move autonomously. It climbs hydraulically and is consistently anchored to the structure using guide rails.

Table 2.1 The most common types of pier construction methods

Material/ Method	Shape	Image
In-situ concrete/traditional formwork piers (photo: SHP, Stráský, Hustý a partneři s.r.o.)	Vertical constant section Vertical variable section	
In-situ concrete/climbing formwork piers (photo: ConstruGomes – Engenharia e Construção S.A.)		
In-situ concrete/sliding formwork piers (photo: ConstruGomes – Engenharia e Construção S.A.)		
Precast concrete piers (any location) (photo: Precast concrete piers for bridge | precast bridge elements; quantity-takeoff.com)		
Precast concrete piers by segments (any location) (photo: ABC, Accelerated bridge construction techniques for the Samuel de Champlain Bridge)		
Steel piers (photo: Nicolas Janberg / Structurae)		
Precast or in-situ concrete V-shaped piers (photo: the civil engineering website)	V-shaped constant V-shaped variable section	
Concrete or steel V-shaped piers in basculation (photo: CFCSL, Carlos Fernandez Casado, S.L.)		

2.2.2.1 Crane Climbing Formwork

Crane climbing formwork is a versatile system suitable for a variety of structures. It facilitates concreting in lifts and allows for the construction of panels up to 7 metres high. The formwork is securely connected to the scaffolding platform of climber units, which are raised by crane to the next phase. This lifting process also transports work platforms for subsequent tasks, such as setting and removing forms, reinforcing, and concreting.

Leading tie rods, embedded in concrete, are affixed at the top of the formwork. Anchorage in the previously hardened concrete section of the pier transfers the lifting device's bearing forces into the structure. The number of concreting sections that can be constructed is limited only by the crane's height

Girder wall and frame formworks are commonly employed in constructing piers with crane climbing formwork. Depending on the wind load, the lifts' height, and the pier's shape, typically one platform or formwork element on each side is sufficient.

Piers are usually segmented into concreting lifts of approximately 3.50 to 5.00 metres in height. It is essential to ensure that the lifting device and formwork are economically adaptable to any variations in the cross-sections of the pylon shaft or pier head [2].

Fig. 2.9 Self-climbing formwork (photos: ConstruGomes – Engenharia e Construção S.A.)

2.2.2.2 Self-Climbing Formwork

Self-climbing technology (Fig. 2.9) is particularly beneficial for constructing very high piers, becoming economical typically after ten phases. The climbing support is equipped with a hydraulic device that allows autonomous movement. This allows the platform, along with the formwork, to be advanced to the next phase without the need for a crane, even under very windy conditions.

Reinforcement work and cleaning of the formwork surface can be performed protected by the retracted formwork. For varying pier cross-sections, it is crucial to adjust the formwork progressively. Planning must account for potentially time-consuming reconstruction efforts to prevent construction delays.

When constructing inclined piers, a portion of the concrete weight acts as an additional load on the formwork, which is a significant factor in the structural analysis of the shoring [2].

The assembly process for climbing formwork includes:

- Constructing the climbing formwork from steel members
- Attaching concrete form panels to the frame, with some supported on rollers
- After pouring a concrete section, the formwork is released and moved away from the pier
- Elevating the frame to the next level using jacks, after concrete in the current lift achieves the required strength
- Once in position, the climbing formwork panels are closed
- Pouring the next concrete section
- This cycle is repeated every three to five days.

A self-climbing rigid steel structure typically comprises three work platforms (top, middle, and bottom). Using six or seven platforms can yield higher formwork speeds compared to those achieved with crane-climbing formwork.

In summary, self-climbing formwork is highly versatile in the construction of bridge piers, accommodating various shapes and heights. Its independence from cranes, flexibility, cost-efficiency, and superior safety features are particularly noteworthy.

2.2.3 Sliding or Gliding Formwork

Sliding formwork (Fig. 2.10) involves pouring concrete into the top of a continuously moving formwork. As the concrete is poured, the formwork is raised vertically at a speed that allows the concrete to harden before it is free from the formwork at the bottom. The concrete used needs to be workable enough to be placed into the form and consolidated by vibration, yet quick-setting enough to gain strength as it emerges from the form. The setting rate of the concrete must be constantly monitored to ensure that the forms are not being raised too quickly. This method is a self-contained formwork system and requires little crane time during construction. It can be used to form any regular shape or core. The formwork rises continuously at a rate of about 20 to 30 cm per hour, supporting itself on the core and not relying on support or access from other parts of the building or permanent works. [3]

At the base of the structure to be built by sliding, a small formwork (about 1.25m) is mounted, forming the beginning of the walls or piers to be executed. Rods or tubes are placed approximately at the axis of the walls, depending on the robustness required by the work to be performed. These serve to fix the jacks that move the formwork, supporting the structure through the fixing heads of the hydraulic system. The formwork's support structure is typically made up of bars or strands spaced 1.5 to 2.5m apart. It is on these supports that the lifting devices will be assembled, which consist of two hydraulic cylinders and two clamping heads which only allow for upward movement. All the devices are connected to each other and to a command centre where the hydraulic pumps are located. The pressure in these circuits is typically between 170 and 200 bar.

The sliding speed depends solely on the setting time of the concrete and, in normal execution situations through small increments of about 6 cm, it can reach values of around 24 cm per hour. Each lift is per-

Fig. 2.10 Sliding formwork (photos: ConstruGomes – Engenharia e Construção S.A.)

formed rapidly (2–3 minutes), with careful control over the height and levelling of the formwork. The lifting devices rest on risers which are placed in cavities defined by the sheaths in the cast concrete, resting laterally in the areas where resistance is guaranteed by the hardening of the concrete. The tubes may be recovered during or at the end of the sliding.

2.2.3.1 Production and Laying of Concrete and Reinforcement

Reinforcement should be prepared as early as possible to prevent limiting glide velocity; however, lapping often needs to be performed in situ. Whenever feasible, bars should be transported to the upper or intermediate platform in anticipation of their use.

Concrete should be laid in thin layers, maintaining a level approximately 5 cm below the top edge of the formwork. The recommended thickness for these layers is between 25 and 30 cm for the following reasons:

- To ensure the monolithic integrity of the pier or wall
- To minimise the frictional force exerted by the concrete on the formwork
- To allow the concrete to achieve sufficient strength, ensuring that no deformation occurs in the freshly laid concrete.

The setting of the concrete should commence within 2 to 7.5 hours and conclude 3.5 to 12 hours after its production. This time frame is determined by factors such as the concrete transport time, the sliding speed, and the ambient temperature.

2.2.3.2 Concrete Vibration

Vibration should be carried out using vibrating needles that are appropriately sized for the thickness of the concrete wall and the spacing between the reinforcements. The use of formwork vibrators should be avoided.

2.2.3.3 Slide Speed

The minimum sliding speed is determined by the requirement that the concrete does not adhere to the formwork, while the maximum speed is constrained by the need to ensure the non-deformability of the freshly poured concrete. Consequently, temperature is a critical factor in determining the number of jack strokes per hour.

Another important consideration is the organisation of multiple work crews, given that this process requires continuous operation, which will be conducted in shifts. Sliding speeds approaching 20 cm per hour are commonly attained with this method.

2.2.4 Precast Piers

Prefabrication for bridge piers construction is an alternative to using traditional formwork systems, climbing formwork, or sliding formwork. The main advantages of prefabrication are related to the industrialisation of the construction processes, improvements in concrete finishes and overall quality, and shorter execution times. This type of prefabricated construction has evolved considerably, and nowadays it can be adapted to the construction of many types of piers, from those for small overpasses to those for large bridges with piers over 40 metres high. Since increasingly powerful auxiliary transport and assembly systems are available, they allow handling weights from 100 to 200 tonnes, although these figures can be far exceeded.

The usual types of prefabricated piers are as follows:

- Independent shafts with or without a supporting column head
- Portal piers made up of vertical shafts and a cap beam
- Piers made up of horizontal segments.

One of the challenges of prefabrication is the connection between precast elements or between precast elements and parts "in situ", especially in hyperstatic structures. Small column sections (about 60×60 cm^2) are usually embedded in pockets left in the foundation footings and the free void is filled with concrete. However, for larger sections, 100 mm diameter corrugated sheaths are usually left in the footing, long enough to anchor the shaft reinforcement. These sheaths are then filled with a non-shrink mortar.

The assembly of these prefabricated elements begins with soft wooden supports that wedge the pieces and prevent stress concentrations on the joint surface. These joints are then filled and adjusted with a liquid mortar without shrinkage to guarantee the transmission of stresses.

For high piers, hollow sections are usually adopted to optimise the use of material, reduce the weight, and facilitate transportation and assembly. Octagonal or I-section piers linked two by two to form a box section are also commonly used.

Further information can be found in [4] and it is summarised in the excerpt below:

> "Nowadays a variety of precast column systems are available in bridge construction. These systems are typically classified as full-height precast columns, precast segmental columns, and precast shells. Also, a variety of connection details, including grouted ducts, grouted splice couplers, post-tensioned joints, socket connections and pocket connections are commonly used. The column systems and connection details can be evaluated by considering aspects such as fabrication, construction, and durability".

2.2.4.1 Precast Reinforced Concrete Columns

This system comprises a full-height precast column element that is designed and detailed like conventional cast-in-place reinforced concrete columns (Fig. 2.11). They are generally used on overpasses or bridges of low height, generally not exceeding 10 m.

If a single support is placed on the shaft and the section of the shaft is sufficient, it is not necessary to place a bent cap and a pier-deck monolithic connection is required, otherwise a bent cap should be provided. If two supports are required, either longitudinally or transversally, the column head can be opened in the form of a palm tree or a bent cap could be considered.

These shafts can be of very varied cross-sections, e.g., circular, square, or polygonal, according to the required aesthetics.

Connections to foundations and bent caps are typically executed by connecting the column longitudinal bars to the adjacent member using grouted splice couplers or by extending the column longitudinal bars into the adjacent member. These connections must ensure that the bars have sufficient development length within the splice coupler or the embedment region. Another way to connect the precast column to the footing is through a socket connection where the precast column is embedded in the cast-in-place footing prior to concrete casting.

Advantages:

- It is the most practical system in terms of fabrication and construction, taking full advantage of the prefabrication of the elements and minimizing the connections
- It improves quality control and accelerates construction

Fig. 2.11 Precast pier columns (photo: https://www.fhwa.dot.gov)

- Large storage and construction areas might not be needed as the piers could be delivered on the installation day.

Disadvantages (limitations):

- The self-weight and height of the system are limited by the erection equipment capacity. As a result, it is not feasible for large piers
- Connections to foundations and superstructure are critical and shall be carefully analysed
- Lift plans are usually complex and an experienced construction engineering team is required
- Construction tolerances must be increased as much as possible to avoid deviations in the connections.

Lessons learnt:

- Rectilinear cross-section shapes are preferred over curvilinear geometries for fabrication
- Several studies recommend a maximum slenderness ratio of about 100
- The weight of the precast column is limited by the capacity of the lifting equipment. As a general guideline, it is recommended that the weight

of the column should not exceed the weight of the girders so that the same crane can be used.

2.2.4.2 Precast Segmental Columns

This system is efficient for tall bridges where the size of the columns makes the use of full-height elements impractical or even impossible due to limitations in weight and dimensions for their transportation and on-site erection.

Column segment connections can be made by splicing the vertical reinforcement with grouted splice couplers, or by post-tensioning the joints. A common alternative is to combine an "in situ" starter segment on top of the footing or pile cap, followed by precast segments. This alternative allows for increasing the reinforcement in highly stressed areas without interfering with the post-tensioning.

Column segments can be either match-cast or epoxy-grouted between segments.

Advantages:

- It is an efficient solution for tall bridges when the use of a full-height column is not feasible, allowing for transportation and erection with standard equipment
- It provides more flexibility for columns with large cross-sections.

Disadvantages (limitations):

- More work is required to connect the column segments at every level.
- Heavy equipment (SPMTs and cranes) is required
- This solution requires a larger storage area surrounding the final location, depending on the pier height

Lessons Learned / Recommendations:

- Using taller segments can speed up construction, while the segment weight governs dimensions for transportation and installation
- Epoxied match-cast is generally preferred over "dry" match-cast (more efficient assembly, good durability, and crushing of segment edges is avoided)
- Hollow sections may require the bottom segment to be filled with concrete for protection from vehicle collision (on land) or vessel collision (in water).
- It is suggested that a dry fit-up of each connection should be performed in the shop before shipping
- Large construction tolerances are crucial for a high-quality final product. Oversized ducts for PT bars and PT strands are required
- For inclined piers, it is recommended to design the pier to be self-stable during the erection to avoid props or falseworks. Moreover, some codes like AASHTO require zero tension in the joints during construction and in-service which could govern the design of the pier.

2.2.4.3 Precast Concrete Column Shell

This system comprises precast hollow elements with thin walls that serve as permanent formwork of a cast-in-place concrete core (Fig. 2.12). In the traditional system used in the USA, the precast shell element is lowered down over a column reinforcement cage, secured in place, and filled with concrete. A similar system developed in Japan uses precast panels as permanent formwork for short solid piers; for taller piers, the panels are used as structural elements to build hollow sections.

The structural performance of the Japanese system is similar to that of the cast-in-place piers, according to experimental testing.

Advantages:

- Reduced weight
- Allows for larger cross-section size and column heights.

Disadvantages (limitations):

- Grade control is challenging during column erection
- Survey and monitoring control during concreting activities is paramount to avoid deformations.

The traditional American system requires installing a reinforcing cage in the cast-in-place core. Not placing the main reinforcement in the shell leads to a reduced effective depth of the section and does not provide significant time-saving.

Non-shrink concrete is recommended to avoid the reduction of the bond between the fill concrete and the shell.

Joints or connections are one of the critical elements for precast construction. The following technologies and details can be applied to connect a precast column to a footing or cap beam and to connect precast column segments together:

- Grouted splice/sleeve couplers: The connection is executed by splicing the vertical reinforcing bars in grouted coupling devices, which are typically proprietary. Sleeves are first cast in the precast element and then grouted after the connecting reinforcement is inserted into the sleeves during erection

Fig. 2.12 Precast concrete column shell (photo: https://poundfield.com/)

- Mechanical splice connectors: Mechanical couplers are used to connect vertical reinforcing bars extending from the column or adjacent member
- Rebar extension into corrugated ducts: These connections use corrugated ducts installed in the precast element during fabrication. During erection, the vertical bars are inserted into these ducts before grouting is applied
- Rebar extensions into grouted pockets: This type of connection requires voided pockets in the adjacent elements. During erection, rebars extending from the columns are inserted into the pockets before grouting is applied
- Post-tensioned joints: Post-tensioning steel bars are installed in the column elements and adjacent members. The bars are coupled at the joint levels and post-tensioned
- Socket-type connections: This connection is executed by extending the precast column into the footing and casting the reinforced concrete footing around it.

Fig. 2.13 "Delta-Pier"

2.2.5 Other Types of Piers

The previous sections were dedicated to vertical piers. The construction of the other types of piers listed in the instruction is very briefly discussed in the present section.

The construction of all these types of piers has similarities with the precast reinforced or segmental concrete piers but differs in the material chosen (concrete vs steel) or in the complexity of the solution and the means and methods for construction (precast, in situ, or basculation V-shaped piers).

Steel piers have been used in the past for all kinds of bridges, from small overpasses to cable-stayed bridge pylons. Design and fabrication of this type of piers are complex but the weight of the elements makes them competitive against precast concrete solutions.

V-shaped piers and Delta-Piers (Fig. 2.13) are broadly used to reduce span lengths but there are some challenges associated with the size of the pier itself. For small piers, in-situ or precast solutions could be developed on falseworks, before placing the superstructure to stabilize the system. For tall piers, adding a tie between the two elements of the "V" is usually done with strand cables to help reduce the demands due to self-weight on the falsework and to better control the erection of the pier and opening of the elements to form the "V" shape.

Sometimes, an inverted V-shaped pier has been designed similarly to some arches construction as it is a simplification of that bridge typology. For this type of piers, a common solution is to precast or cast in situ both shafts in position (vertically or horizontally) and connect them to a pin in the ground. Then, both shafts are rotated into their final position. Similar to the V-shaped piers or the arch bridge construction, ties with strand cables (to allow rotation) are used as auxiliary elements during construction.

2.2.6 Lessons Learnt and Recommendations

In addressing the common risks and challenges in pier construction, in addition to the notes on the impact of construction equipment on the piers [5], other actions during construction are relevant [6] combined with the influence of geometric deviations on these procedures, as stated in sections 2.4.4 and 3.1.6. Hence, the following issues should be considered:

- Topographic Control: This is crucial, especially when dealing with inserts or hollows in piers, meaning that a double check of topography is highly recommended in these scenarios. Concerning inserts or hollows related to construction equipment fixing, a minor deviation in their execution (for example, tens of millimetres) may have more severe consequences than a large deviation in pier position (such as tens of centimetres)

- Accident Prevention: The most frequent accidents involve failures of anchorages, prestressing bars, or bolts. This is mitigated by implanting redundancy in anchorages and prestressing bars. For bolts, consider inverting the screw threads of pier formwork bolts (reverse tightening direction) to prevent the use of incorrect bolts. These formwork systems require certain concrete strengths to be achieved at an early stage in previous lifts to enable resistance to the working loads and moments resulting from the pressure exerted by the fresh concrete being poured into the lifts
- Working Platforms: Conduct a comprehensive risk analysis, including an assessment of the entire path that the workers take from their arrival to the location of their specific tasks in the pier erection process. A method for emergency evacuation of personnel needs to be planned in the event of structural distress
- An operation manual, the training of personnel, and safety inspection procedures are important for safe operation
- Oscillations in High Piers: Significant oscillations (e.g., due to wind) can occur in very tall piers, sometimes leading to the workers' reluctance to work for extended periods. This can be addressed by providing special training during the hiring process and informing the workers about these conditions
- Concrete Piers Over Water: Procedures for erecting concrete piers over water need careful planning to synchronize transport and pouring times. Using additives or adjusting the volume of transported concrete may be necessary
- Post-Design Construction Method Decisions: If construction methods, including those for deck construction, are decided after the design stage, it is crucial to assess the construction equipment's impact on the piers, as indicated in the subsections referred to above [7].

References

[1] https://constrofacilitator.com/an-overview-of-climbing-formwork/

[2] https://constrofacilitator.com/slip-formwork-application-components-and-advantages/

[3] https://www.perihk.com/dam/jcr:28852915-da9b-4420-91dd-e644bcaabf1f/bridge-formwork-technology-manual.pdf

[4] Synthesis of Precast Bridge Column Designs", Ghassan Fawaz, Juan Murcia-Delso, Oguzhan Bayrak – TECHNICAL REPORT 0-6978-1 (FHWA/TX-19/0-6978-1), Center for Transportation Research at The University of Texas at Austin

[5] AASHTO. Construction Handbook for Bridge Temporary Works, 2nd ed. 2017

[6] European Standard EN 1991-1-6:2005. Eurocode 1: Actions on structures. Part 1–7: General actions – Actions during construction.

[7] "Formwork for Concrete" M.K. Hurd, ACI publication

https://doi.org/10.2749/sp001.ch02.3

2.3 Bearing and Joint Installation

Alp Caner[1] | Niculin Meng[2]
[1] USA, [2] Switzerland

As is well known, bearings and joints are essential in bridge engineering, as they directly influence the definition and flexibility of the structural system and significantly impact the bridge's integrity and durability. Their installation is included in the critical path of project planning and must adhere to technical standards to prevent any adverse structural responses or operational issues, both during construction and in the service stage.

This section is based on European Standard EN1337, VHFL Guideline, Part2 [1, 2] with a view to creating a general guideline to mitigate bearing installation failures that can lead to malfunction of the abovementioned construction products.

2.3.1 Introduction

A detailed bearing schedule, including the corresponding loads, must be provided by the bridge designer. The purpose of a bridge bearing schedule is to list the information normally required for the design of the bearings. This information should ensure that bearings are designed and constructed to prevent myriad actions from giving rise to adverse effects of the bearings on the structure (i.e., maximum horizontal load with minimum vertical load). A drawing should accompany the schedule showing the layout of the bearings with identification marks, including a typical cross-section of the bridge and particulars of any special locating requirements. The anchorage of the bearing should be defined according to the worst-case load combinations. Unanticipated load combinations can cause bearing and/or structural damage (Fig 2.14).

All geometrical requirements (connection to steel/concrete structures) and permissible pressures on structures (steel or concrete grades) should be defined prior to bearing design and coordinated with the bearing manufacturer.

Fig. 2.14 Examples of bearing and structural joins damage (bottom photo: https://www.cbc.ca/news/canada/thunder-bay/nipigon-river-bridge-closed-transcanada-1.3397831, on 10/24/19)

Fig. 2.15 Example of design failure leading to cracks on shear keys

Fig 2.15 shows that the shear keys which are intended to carry only transverse earthquake loads were mistakenly constructed larger than planned and thus carry the vertical loads instead of the bearings. These shear keys are cracked at the expansion bearing locations. Such design details need to be clearly communicated to the construction crew.

2.3.2 Orientation and Pre-Setting

Firstly, the bearing layout and the bearing installation drawing should be obtained from the bridge designer. In addition, orientation and pre-setting information should be provided in the contract plans. A comprehensive list of bearing installation requirements is provided in EN1337-11 (German National Annex) [3] and summarised below:

a. Bearing type (Fig. 2.19), bearing dimensions, and materials used
b. Details of bearing points (i.e., block-outs, plinth, and reinforcement)
c. Fixity details and information on grouting of the bearing
d. Installation heights
e. Inclinations in longitudinal and transverse direction
f. Wedge plate dimensions
g. Geometric tolerance requirements and information on the limit values of climatic temperature effects during installation
h. Information on bearing presetting as a function of the structural temperature during installation of the bearings. (creep and shrinkage of concrete, elastic shortening due to the remaining prestressing may also be mentioned).

Pre-setting information should be given in the bearing drawing provided by the bearing manufacturer. The presetting direction and presetting value (in mm or inches) should also be labelled on the top plate of the bearing by the manufacturer. Furthermore, each bearing is to be marked with the type, design load, weight of the bearing, project number, bearing location, and orientation axes on the top plate of the bearing. An identification plate should also be mounted at the side of the top or bottom plate to allow traceability after the installation of the bearing (Fig 2.16).

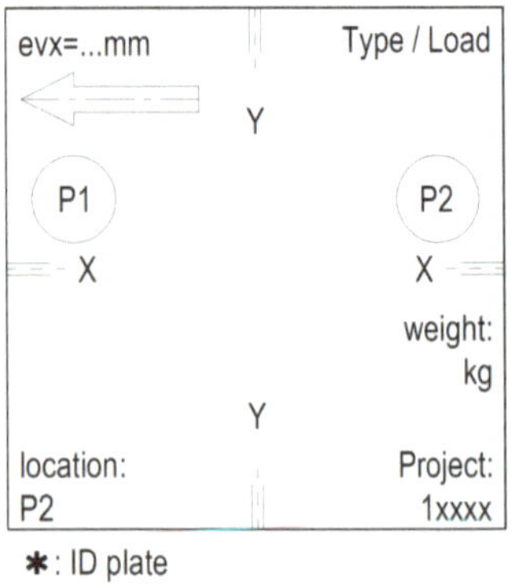

Fig. 2.16 Example of bearing labelling – top plate

Fig. 2.17 ▶ Example of a bearing layout

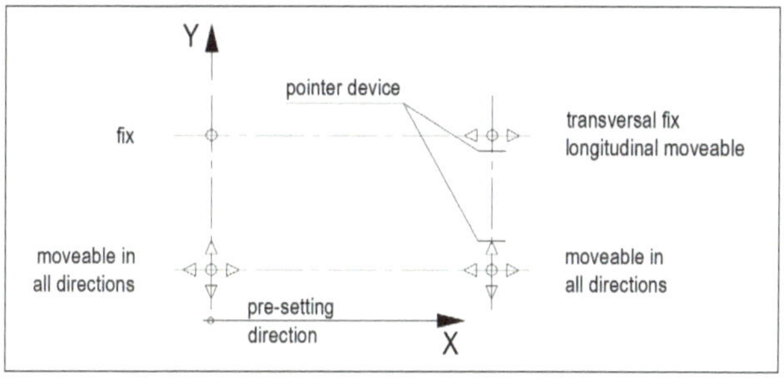

2.3.3 Installation Sequence

It is recommended that the bearings be installed under the supervision of the representative or qualified personnel of the manufacturer following the steps listed below [3]:

- compliance with the bearing layout (Fig. 2.17)
- position of the bearing including alignment, preset, and transport fixing, taking into account the actual temperature of the structure (Fig. 2.18)
- examination and condition of the contact areas
- mixing and placing of joint mortars or other levelling materials
- checking the strength of the mortar
- release of transport clamps (Fig. 2.18)
- providing temporary restraints in relation to erection phases
- zero-measurement / commissioning inspection
- checking for any changes in measuring point position due to movement and loading.

Fig. 2.18 Failure due to wrong orientation and un-released transportation bolts

Symbol	Bearing type
	Steel Spherical bearing/Pot bearing steel point and rocker bearing, fixed in any horizontal direction
	Steel spherical bearing/Pot bearing steel point and rocker bearing with uniaxial movable sliding part
	Steel Spherical bearing/Pot bearing steel point and rocker bearing with biaxial movable sliding part
	Elastomeric bearing with securing device for two axes
	Elastomeric bearing, biaxial deformable, with/without upper and lower bearing plate
	Steel Spherical bearing/Pot bearing steel point and rocker bearing, fixed in any horizontal direction
	Elastomeric bearing with unidirectional movable sliding part and Guide bearing for the other axis
	Elastomeric bearing with multidirectional movable sliding part
	Elastomeric bearing with unidirectional movable sliding part
	Guide bearing with restraints for one axis
	Restrained bearing, fixed in any horizontal direction

Fig. 2.19 Common bearing symbols/types

2.3.4 Levelling

Bearings should be levelled using screws (Fig. 2.20), steel shim plates, or any other suitable temporary supports under the bearing base plates (Fig. 2.21).

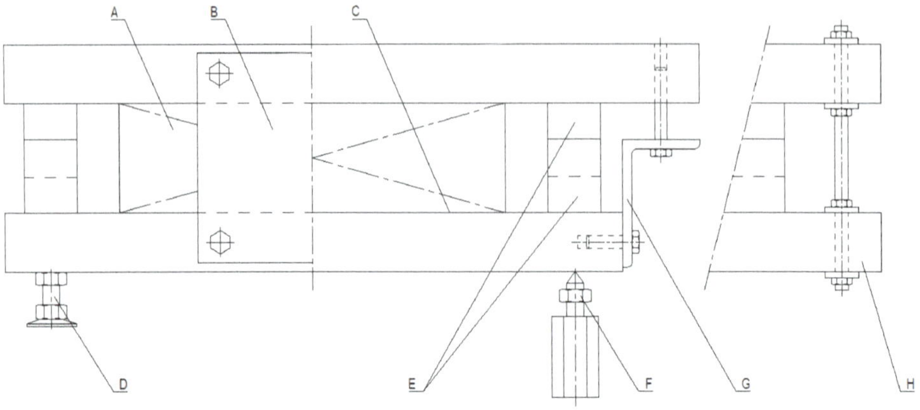

A Elastomeric pad
B Assembly device, variation 1
C Bearing normally installed horizontally (deviation after installation + 0.5%, measured over the diagonal length)
D Levelling screw with compression cone
E Guiding
F Levelling screw, needle center
G Assembly device, variation 2
H Assembly device, variation 3

Fig. 2.20 Levelling example with screws

Fig. 2.21 Levelling and general installation failures

Normally, bearings are installed level, where the measured deviation from the horizontal plane on the x-axis and y-axis should not exceed 0.3 %, or 0.5 % for elastomeric bearings, respectively. Unless otherwise considered in the design of bearings, the bearing should be located so that its position does not deviate more than 3 mm from its nominal position in any direction. For continuous girders, special attention should be given to the influence of deviations in the vertical direction, which can cause large deviations in support reactions.

2.3.5 Grouting Works

The following methods are suitable for the grouting of bearings (this part is mainly taken from EN1337-1 [4]) (Fig. 2.21):

- placing the bearing on a cambered bed of plastic mortar in such a way that excess mortar can be squeezed out on all sides; or
- placing it on a thin layer of levelling mortar; or
- bedding by grouting, using free-flowing mortar. This is preferable

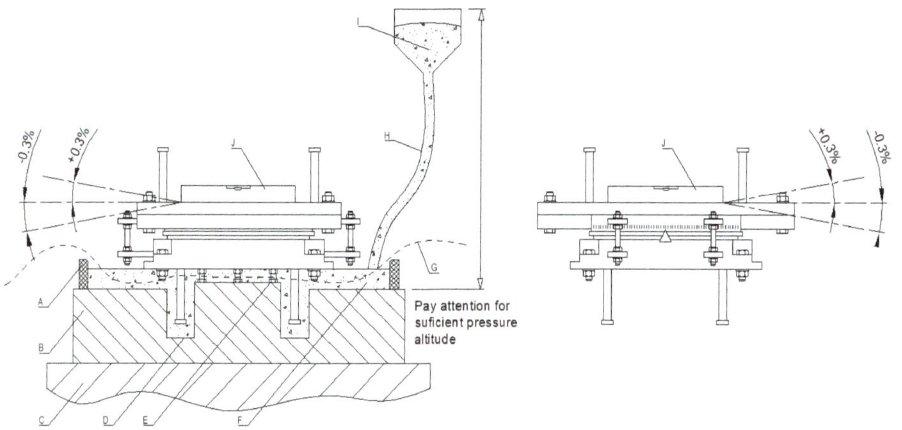

A	Casting	F	Mortar joint
B	Plinth	G	Chain
C	Substructure	H	Plastic hose
D	Box out for anchor, dolls or studs	I	Liquid mortar
E	Levelling bolts for proper adjustment and levelling	J	Levelling by a spirit-level in both directions

Fig. 2.22 Grouting and levelling example

for bearings with headed studs. The formation of air bubbles in the mortar must be prevented. For this, metal chains are inserted parallel at a distance of 15–20 cm, and pulled back and forth, releasing the air; or

- bedding by packing mortar beneath the bearing; this method is only recommended for use where the smaller side length is less than 500 mm.

Fig. 2.23 Non-shrink grout cracking

Bearings that are to be installed on temporary supports should be firmly fixed to the substructure to prevent disturbance during subsequent operations. The method of tightening should be done in a manner that does not deform the bearings. Finally, voids beneath the bearings should be completely filled with bedding material using the appropriate method.

This can be achieved by, for example, removing temporary packing shims and using resilient washers under nuts. Alternatively, bearings may be fixed directly to anchor plates that can be cast in or bedded on top of the supporting structure to the correct level and location.

The maximum layer thickness should not exceed 50 mm for cement mortar and 30 mm for epoxy resin or methacrylate mortar.

In the case of cement mortar or cement grout, the plinth concrete surface should be saturated with water before installation to prevent dehydration. Any remaining water on the surface should be blown out immediately before the mortar is poured.

	0			1	2	3	4
1	Installation location (support no./alignment) as per drawing						
2		Bearing type (abbreviation according to prEN 1337-1) / Bearing no.					
3		Load F_z in kilonewtons					
4		Horizontal forces F_x / F_y in kilonewtons					
5		Calculated displacement in millimetres, + indicating the direction away from the fixed point	V_x ±				
			V_y ±				
6		Setting in millimetres	e_{xv} ±				
			e_{yv} ±				
7		Drawing no. / sheet no.					
8		Date of delivery					
9	Prior to installation	Correctly unloaded, laid on squared timber and covered					
10		Location of marking on upper face of bearing					
11		Indicator provided					
12		Plate indicating type of bearing					
13		3-pin measuring plane on lower face of bearing					
14		Cleanness and protection against corrosion					
15		Correct and firm seating of temporary clamping devices					
16		Installation location as per line 1					
17		Raising of superstructure Date and time					
18		Cleanness of mortar contact surfaces					
19		Direction and magnitude of presetting in millimetres, + indicating the direction away from the fixed point					
20		Deviation from horizontal, in millimetres per metre, determined at the measuring points (longitudinal/transverse)					
21	Installation	Placing of mortar	Date				
			time (from … to …)				
22		Air temperature / temperature in structure, in degrees Celsius					
23		Thickness of mortar joints in millimetres (u) = unreinforced (b) = reinforced	upper				
			lower				
24		Lowering of superstructure Date / time					
25	Start of function	Temporary clamping devices released / removed					
26		Protection of sliding surfaces in place					
27		Cleanness and protection against corrosion					
28		Date / Time					
29		Air temperature / Temperature in structure, in degrees Celsius					
30	Initial measurements	Deviation from horizontal, in millimetres per metre, determined at the measuring points (longitudinal/transverse)					
31		Displacement in mm, + indicating the direction away from the fixed point	Vx / Vy				
32		Sliding clearance in mm	max / min				
33		Tilting clearance in mm	max / min				
34	Remarks or other information, e.g. with regard to erection procedures, temporary alterations of fixed points, etc.						
Remark: Bearings shall only be adjusted by means of the levelling screws.							

Fig. 2.24 Bearing Installation report (excerpted EN1337-11;1997)

When resin-based mortar is employed, the chemical characteristics of the resin and the resin/filler ratio should ensure a satisfactory consistency and setting time to ensure the correct installation under site conditions. Durability in terms of strength, final hardening, and shape should be considered.

If resin-based mortar is expected to be in direct contact with the bearing, the chemical compatibility and friction coefficient should be verified by testing unless prior satisfactory use in similar conditions can be demonstrated.

The thickness of mortar between the bearing and the plinth should not exceed the limitations given by the mortar manufacturer. Special attention should be given to local climate conditions.

Removable formwork should remain in place until the mortar has hardened sufficiently but should be completely removed by the time the bearing is fully operative. Removal by burning off is not permitted.

The weight of the superstructure should not be transferred onto the bearings until the bedding material has developed sufficient strength to resist the applied load. Temporary clamps should be removed before the bearings are expected to accommodate movement. Consideration should be given to any treatment needed for holes exposed upon the removal of temporary transit clamps. Where reuse of these fixing holes might be necessary, the selected filling material should not only offer protection against deterioration but should also be easily removable without damaging any threads.

The non-shrink grout used between the pedestal and anchor plate helps to ensure full contact between bearing and grout (Fig 2.21). For larger bearings, the grout can crack due to its non-shrink properties as the concrete below is free to shrink over time, although the cracks do not usually extend into the reinforced concrete pedestal. The material composition and thickness of the grout can be adjusted to overcome this problem

2.3.6 Start of Bearing Operation

After the formwork is removed, all the bearings should be cleaned and any damage to the corrosion protection system should be repaired. Working with fire, flammable goods, or chemicals near the bridge bearings should not be permitted. A bearing installation report (Fig. 2.24) including the initial measurements should be completed before, during, and after installation.

2.3.7 Lessons Learnt and Recommendations

- In bridge construction, engineers often delay the decision to select bearings until a later stage. These high-standard products require a considerable amount of time for manufacturing and quality control. Frequently, underestimating the delivery time leads to production delays, resulting in instances where the construction of the bridge deck is on hold, pending the arrival of the bearings.
- During the preparation of construction stage drawings, it is crucial to ensure that the geometry of the bearings is compatible with the supports for the deck equipment. Although the geometric specifications of bearings are typically con-

firmed during the design stage, conducting an additional verification is advisable to prevent unexpected issues and interruptions during production onsite.

- Provisions for the replacement of bearings in the future need to be considered right from the design stage and should be shown in the related drawings

References

[1] EN 1337-2, Structural bearings – Part 2: Sliding elements.
[2] VHFL RiLi Leistungsumfang des Lagerherstellers bei der Einbauaufsicht Lager.
[3] EN 1337-11 Structural bearings – Part 11: Transport, storage und installation.
[4] EN 1337-1 Structural bearings – Part 1: General design rules.

2.4 Essentials on Deck Construction

Renato Bastos
Portugal

2.4.1 Introduction

In bridge engineering, the choice of the deck construction method is a central aspect of any design solution. This chapter provides an overview of the most common construction methods, both for concrete and steel, whether cast in situ or precast. The key characteristics of the most frequently used methods are highlighted. Additional information on bridge construction methods is provided in Sub-chapters 5.1 to 5.4, where different types of deck construction equipment are discussed. To provide clarity, prior to the deck methods descriptions, some basic definitions are herein presented.

2.4.1.1 Definitions

The related literature review [1–18] revealed conflicting use of the terms shoring, formwork, and scaffolding. The definitions set forth below are adopted to avoid confusion with the terminology used in this section. These definitions are not intended to be exclusive but are generally consistent with the common use of these terms.

Falsework – Any temporary construction work used to support the permanent structure until it becomes self-supporting. Falsework covers steel or timber beams, girders, columns, piles and foundations, and any proprietary equipment including modular shoring frames, post shores, and adjustable horizontal shoring.

Formwork – A temporary structure or mould used to retain the plastic or fluid concrete in its intended shape until it hardens. Formwork must have enough strength to resist the fluid pressure exerted by plastic concrete and any additional fluid pressure effects generated by vibration.

Scaffolding – An elevated work platform used to support workers, materials, and equipment, but not intended to support the structure.

Shoring – This refers to a falsework component such as horizontal, vertical, or inclined support members. However, for the purposes herein this term is used interchangeably with falsework.

2.4.2 Concrete Decks

2.4.2.1 Cast-in-situ Span-by-Span with Static Falsework

Not disregarding other possible adequate solutions, this type of falsework is advisable for the construction of viaducts with small spans (up to about 40 metres) and moderate height (up to about 25 metres) on stable and regular terrain and, generally, where there is no need to keep free distances for any type of crossing. The spans need not be equal, nor does the deck's cross-section need to be constant [6]. Certainly, the appropriate decision should be evaluated on a case-by-case basis, taking into account the characteristics of each possible solution.

The construction time, using this type of falsework and for the usual deck widths, i.e., between 11 and 16 metres, is about 2 to 3 weeks per span, for usual span lengths of around 35 metres.

Floor-mounted falsework is essentially of two types:

- tubular structures forming continuous supports in which the main internal forces on the shoring elements are axial (classical tubular shoring in Fig. 2.25) or
- structures with shoring towers and truss beams, allowing, locally, for free spans underneath, in which the main internal forces are axial (towers) and bending (trusses).

2.4.2.1.1 Classic Tubular Shoring

These shoring systems are simply made up of tubular elements (circular hollow sections) mounted vertically on ground supports (e.g., wooden planks, sleepers, or concrete blocks), braced by other tubular elements mounted vertically and diagonally, and linked together by steel clamps. Height levelling is done by inserting wedges and hardwood pallets either at the lower end of the tubes, between them and the ground support, or at the upper end, between the tube and the deck formwork.

This is a simple system to assemble and it fits perfectly in the case of small works. However, when it was introduced into the market, this type of falsework had spectacular applications in the construction of projects with a large span and height, before being relegated to more modest uses following the emergence of newer technologies.

2.4.2.1.2 Technological Developments of Classic Tubular Shoring

Using classic tubular shoring as an inspiration, different systems were developed with a view to simplify the erection and dismantling operations, primarily by enabling multiple cycles of assembly and dismantlement and the easy recover of the elements after each use.

The widespread use of this kind of system is due to its dismountable nature, lightness, durability, and safety.

The system is quick to use and handle, with very robust connections. Given the absence of screws, it allows for a considerable reduction in the qualitative and quantitative requirement of labour, with its main feature being economy in procurement and maintenance.

2.4.2.1.3 Shoring Towers and Truss Beams

The "tower and truss" system comprises a non-continuous falsework that allows gaps below the bridge deck with a certain clearance, for example, allowing traffic underneath. In some cases, this can be an important advantage for the management of traffic during construction.

It is a system made up of beam and tower elements. The beam elements are trusses made of steel sections, generally of high-strength quality. Beam dimensions can vary enormously up to an economically viable maximum of 35 to 40 metres, the largest of which are obviously heavier.

These beams are supported on towers, consisting of 4 vertical tubes connected to each other, generally also made of high-strength steel. These column elements have different modules up to 3 metres long so that the towers can be built with the required heights.

The towers have adjustment and dismantlement devices at both ends. The process of mounting a falsework of this type essentially depends on the existing lifting systems available.

The main advantage of this system is enabling the passage of traffic or construction equipment underneath. Its drawbacks include the need for a higher ground bearing capacity , sometimes requiring deep foundations for the towers.

Fig. 2.25 Floor-mounted falsework (continuous tubular structure – detail)

2.4.2.1.4 Moving Falsework

When the length of a viaduct exceeds 300 metres and the outer dimensions of the deck cross-section are constant, a floor-standing movable falsework may be a cost-effective solution, provided that the height of the deck relative to the ground is not excessive and the terrain is relatively flat.

The most traditional way of operating a floor-mounted moving falsework is to move it on extendable rails, where the formwork is directly attached to the falsework so that they move together.

During concreting it is necessary to raise the falsework to leave the rails unloaded. After concreting and post-tensioning a particular span, the falsework is then placed back on the rails for support while simultaneously releasing the formwork.

2.4.2.2 Cast-in-Situ Span-by-Span with Movable Scaffolding System

Bridge construction based on this method uses a self-supporting steel launching beam (Fig. 2.26) sustained on the final supporting elements of the bridge (columns and abutments) and, sometimes, one part of the deck that has already been built, moving from one span to the next span autonomously. This procedure makes straddling large valleys, rivers, and urban environments possible without the need to use ground shoring, thus being independent of the constraints imposed by the terrain features and obstacles and keeping free part or all of the area below the deck.

The movable scaffolding system (MSS) consists of a set of longitudinal beams and an auxiliary platform that supports the formwork (or, in particular cases, partial precast elements). The MSS may be classified as an underslung or an overhead travelling gantry, depending on whether the supporting beam is above or below the auxiliary platform, or even mixed, using both upper and lower supporting beams.

The longitudinal supporting elements can be solid beams or trusses and are always constructed to be easily assembled and disassembled for possible reuse.

Fig. 2.26 Movable scaffolding system

2.4.2.2.1 *Field of Application*

This construction process is economically applicable to works of rectilinear axis or small curvature, of long length, and usually of constant section. This method is often used with spans between 30 and 55 metres, with the optimum span between 40 and 50 metres. The length of each concreting phase is normally equal to the length of each span, with the joints being located near the bending moment inversion sections (usually L/4 to L/5). Since all the equipment is independent of the ground, the execution time and the cost of the deck do not depend on the height of the viaduct, as access to the work front is made through the already-built deck, both for people and materials.

Up to 2016, the in-situ, span-by-span, prestressed concrete deck construction of multi-span bridges was limited to spans up to 70 or 80 metres. Following recent scientific and technical validation of the organic prestressing technology applied in movable scaffolding systems, and evaluation of its potential features, new construction limits are now reached, overcoming the former critical range of spans for multi-span viaducts of 70 to 100 metres [8].

2.4.2.2.2 *Advantages*

i) Speed of execution: the construction of a span takes about one week, and it may even be possible to construct each span within 5 days if precast elements are used
ii) Ease of construction of the deck, due to the ease of access to the work front (especially in the case of lower launching beams)
iii) Easy systematisation of operations and the possibility of adjustments and adaptations during the work
iv) Independence of work from the ground, which provides for uninterrupted work due to floods, traffic, etc., and, on the other hand, existing buildings or activities that may take place under the deck are not jeopardised
v) Safety of workers, as they have a relatively large work platform.

2.4.2.2.3 *Disadvantages of the Process*

The main disadvantage of the method is the large investment required for the acquisition, transportation, and assembly/disassembly operations of the movable scaffolding system. These steel structures are very heavy (250 to 500 kg per tonne of supported concrete) as they have to straddle considerable spans by supporting their own weight as well as the concrete weight of one span. The payback of equipment like this is only possible in long viaducts or through their reuse in several works. [7]

Another important concern of this method is wind during construction. Due to the dimensions of the movable scaffolding system, certain wind combinations can influence the foundation design for long and tall structures.

2.4.2.2.4 Overall Concept and Design

The use of this construction method requires the interlinked study of the geometry of the deck, columns, and movable scaffolding system. The deck's cross-section is usually made of T-beams or box-girder beams. The box section is more commonly employed when using precast elements or for in-situ constructions when the spans are over 45 metres, as this type of cross-section can present extra difficulties when the deck is concreted onsite due to the need for interior formwork. The optimum ratio between the height of the deck and the span of each section is approximately 1/15, usually varying between 1/12 and 1/18.

Unlike the incremental launching method and the balanced cantilever method, the prestress of bridge decks built using MSS is practically independent of the construction process and, as such, the deck girder can be designed as if it were a continuous beam.

2.4.2.3 Balanced Cantilever Method (Cast-in-Situ and Precast)

Free cantilevering is a method of construction where a structure is built outward from a fixed point to form a cantilever structure, without temporary support, using a staged cast-in-situ construction. When segments of two arms on either side of a pier are cast symmetrically, it is known as a 'balanced cantilever' (Fig. 2.27).

Basically, cast-in-situ construction describes a process whereby segments are progressively cast on site in their final positions within the structure. Consequently, in this method, the construction of the deck is carried out through segments built successively, cantilevering from its supporting elements.

Fig. 2.27 Balanced cantilever construction (set of 2 twin bridges)

Therefore, each segment is connected to the previous one with sufficient strength to be self-supporting and to support the following segments, i.e., it must bear the actual weight of the segments to be constructed, the formwork and the equipment required for the construction of each segment.

Generally, the construction is carried out building symmetrical precast or in-situ concrete segments on each side of the pier to prevent introducing bending moments into the columns and foundations. After the construction of each segment and before stripping it is necessary to apply the prestress to ensure its integration with the part that has already been built. The concreting or placement of the segments is not simultaneous, however, so the cantilevers are unbalanced for some time. The required balance is achieved by providing continuity between the deck and the pier or by using temporary external connections, for example, prestress cables attached to the foundation massifs (Fig. 2.28). It is also necessary to consider a possible difference in weight of symmetric segments, effects of wind, etc.

Fig. 2.28 Balanced cantilever construction with temporary external cables

2.4.2.3.1 *Field of Application*

The balanced cantilever method is often appropriate and cost-effective for the construction of long-span concrete bridges where height, topography, or geotechnical conditions render the use of conventional formwork uneconomical.

Furthermore, this method is easily adaptable to irregular and long-span lengths, congested project sites, rough terrain and water, rail crossings, and environmentally sensitive areas. [4]

The economical range of span-lengths for the cast-in-situ cantilever construction is normally between 60 and 200 metres, with the most frequently used spans being between 80 and 120 metres. Below 70 metres, this method begins to lose cost-effectiveness in relation to other approaches. For spans greater than 90 metres, it is usually the only available or possible solution.

2.4.2.3.2 *Advantages*

The main advantage of this process is the absence of scaffolding and shoring, thus freeing all space under the bridge. This process is particularly useful for viaducts with high columns, crossing deep and wide valleys where falsework is costly, and for bridges over rivers with strong and variable flows where shoring can be dangerous as well as costly. It is also sometimes used in urban areas, where the clearance below the deck soffit is low and a movable scaffolding system is impractical due to the large size of its constituent elements. There are other advantages such as the use of few formwork elements and their reuse throughout the worksite, good manpower performance due to the mechanisation of the process, and the possibility of accelerating the construction by having several working fronts [7].

2.4.2.3.3 *Disadvantages*

Due to the degree of challenges inherent to this construction process, one of the main disadvantages is the considerable technical capacity required of the contractor responsible for the work. The necessary precision of the formwork advance operation and the accuracy required in the geometrical control of the construction are two good examples of the need for high-level preparation.

For precast bridges of sufficient length, a precast yard for the production and storage of the segments is necessary but usually justified cost-wise. For the design of this facility, it is important to study the storage area, the accesses for delivery of the segments to the site, geometric control during production, the formworks, the equipment required for the production, the concrete plant, and an approved concrete mix.

2.4.2.3.4 *Overall Concept and Design*

Because of the construction sequence, there are hogging bending moments along almost the entire length, being of considerable value near the supporting elements, so that the box cross-section is best suited to these internal forces due to the presence of a wide lower flange. Possible problems concerning the lack of strength of the lower flange can be avoided by increasing its thickness. There are other advantages to using this type of cross-section, such as its high torsion resistance and improved mechanical performance, which means lower stresses, thereby reducing the effects of creep.

On the supports and in the sections of the expansion joints it is usual to use diaphragms for the distribution of shear forces and torsion moments. Openings must be provided in these elements to permit the circulation of maintenance personnel and the passage of pipes (utilities, drainage, and other non-structural conduits).

For roadway bridges, the ratio between the cross-section height over the columns and the span length is between 1/16 and 1/20 for variable height decks and 1/18 and 1/25 for constant height decks, the optimum values being 1/17 and 1/20, respectively. In the case of a variable cross-section, the slenderness ratio of the deck in the mid-span is between 1/35 and 1/60 depending on the deck width and the span length.

Currently, the most commonly used solution is the continuous frame, in which the continuity between cantilevers is achieved through joining their ends by concreting the so-called closing segment and by using connection prestress cables. The use of this static scheme has several advantages, such as the elimination of expansion joints (construction and maintenance benefits), increased safety (stress redistribution), decreased elastic and time-dependent deformations, improved aesthetic appearance, and increased traffic comfort due to the absence of discontinuities in the longitudinal profile of the pavement.

In bridges built by the free cantilever method, the prestress is usually composed of two cable families: cantilever cables and continuity cables. The cantilever cables are laid out near the top slab of the deck and are placed and tensioned throughout the construction process to resist increased hogging moments. The profile of these cables can be inclined along the webs, resulting in the reduction of the shear forces due to the vertical components of the prestress, or can be straight along the upper slab, reducing the frictional losses of the cables and facilitating their threading. The continuity cables are placed in the lower slab of the deck or at the bottom of the webs, near the mid-span, serving to establish continuity between the cantilevers and resist the sagging moments arising from time-dependent effects, the superimposed dead loads, and live loads.

2.4.2.3.5 *Balanced Cantilever Erection of Precast Segmental Bridges*

Nowadays this variant is widely used due to the improvements introduced in the prefabrication and transportation processes, in addition to its low labour costs. The main advantage of the prefabrication of segments is higher construction speed, achieving values of approximately six segments per day and per item of equipment. Other advantages of this method include independence from weather conditions in concreting operations and more efficient control of concrete characteristics, avoiding the difficulties of applying prestress in young concrete, as well as decreasing the effects of concrete shrinkage and creep (as the segments are already a few weeks old when they are installed).

2.4.2.4 Incremental Launching

Incremental launching is a well-developed construction procedure for long girder bridges of a constant depth. It has many advantages including minimal interference with the landscape, the concentration of most construction work at a well-protected and secure location, the industrialisation of the construction work and the corresponding high precision of the resulting structure, and, finally, higher standards of health and safety conditions which can be achieved with this method. The drawbacks are mainly related to the limitations of this method since it requires specific horizontal and vertical alignment conditions as well as a substantial space behind one of the abutments in which to concentrate most of the construction operations. This method has been applied to build numerous girder bridges, especially railroad bridges, but also some bowstring arches.

This method consists of the construction of successive superstructure segments in a preassembly and launching area behind one of the abutments, oriented along the axis of the viaduct.

After the concrete has cured, the elements are then placed in their definitive position by longitudinal translation. In this process, the deck is moved by sliding over its columns, abutments, and some temporary supports (if needed) by the action of hydraulic jacks attached to one of the supports. As appropriate, the deck can be pulled or pushed.

The incremental launching method is particularly suited to the construction of continuous post-tensioned multi-span bridges. It involves casting, typically 15 to 30-metre-long pieces of the bridge superstructure in a stationary formwork behind an abutment and pushing a completed section forward with jacks or a friction launching system along the bridge axis. The sections are cast contiguously and then stressed together. The superstructure is launched over temporary sliding bearings on the piers.

Due to the alternating stresses resulting from successive changes in the support conditions, this method was first used for the construction of steel bridges. The first application of the method to concrete structures took place on the bridge over the Caroni River in Venezuela in 1962/63, designed by Professor Fritz Leonhardt, and is still being used in this type of structure. [7]

The current use of the incremental launching method was driven by the discovery of new materials like polytetrafluoroethylene (PTFE), which allowed the construction of low-friction bearing devices, and by the development of prestress, which made it possible to lighten the structures and facilitate the connection between elements.

To minimise the hogging bending moment in the superstructure during construction, a steel launching nose is attached to the front of the bridge superstructure. As an alternative, a king post could be used to raise the tip of the bridge while passing over the piers.

When the deck is in its ultimate position, the final post-tensioning is installed and the temporary prestress is transformed or eliminated.

2.4.2.4.1 Field of Application

In most cases, this construction process is only possible for bridges or viaducts with a constant deck height and a straight longitudinal profile or constant (horizontal and vertical) curvature. It can also be performed on works with low plan curvature by applying lateral displacements when shifting along the axis of the deck.

The span length of multi-span bridges constructed by this process is usually 40 to 50 metres. The central span can reach 70 to 80 metres when it is larger than the rest and the deck is pushed from both ends. In multi-span and large-span bridges, temporary auxiliary supports may be used to reduce the cantilever span during launching.

2.4.2.4.2 *Advantages*

i) The entire area under the deck is free due to the absence of falsework and formwork, and there is a lower risk of falling parts or materials or reduced headroom (nevertheless, there is always some risk of falling material, as there will be access platforms at the top of each pier to operate the launching system)
ii) Ease of access to restricted or limited sites – such as over rivers, deep valleys, roads or train lines, in poor soil conditions, or environmentally protected areas
iii) With almost all deck construction activity concentrated in a small, easily accessible area, it can be covered, thereby protecting workers from unfavourable weather conditions and ensuring the safety and manufacturing quality that would be achieved in a prefabrication industrial facility
iv) The launching subsystems (whole equipment parts) are light, with moderate costs and little dependent on the characteristics of each deck, so they can be reused in other viaducts, enabling better payback of its cost
v) Since the formwork is reusable and stationary, there are strong cost reductions due to its reuse and the absence of cranes and launching beams for the construction of the deck (this needs evaluation on a case-by-case basis)
vi) Elimination of transportation costs associated with prefabrication systems
vii) Easy geometric control of the deck geometry after the launching operation is concluded (see note below on geometric control during launching operation)
viii) Elimination of concrete joints (there are 2 possibilities of joints: dry joints or joints with resins)
ix) Possibility of shortening lead time by using two manufacturing areas.

2.4.2.4.3 *Disadvantages*

i) The longitudinal profile and the road layout geometry depend on the construction process, sometimes impairing the aesthetic aspect, particularly regarding the achieved slenderness
ii) It is advisable that the height of the cross-section be constant, which is uneconomical for spans greater than 60 metres
iii) A considerable area behind one of the abutments is needed, at least equal to the span length, which is not always available, thus making the use of the process unfeasible
iv) The use of a considerable temporary prestress is quite expensive, despite the possibility of incorporation into the final prestress
v) Strict execution control is necessary, particularly on the face of the superstructure sliding on the supports, and sliding operations (translational force, lateral displacements, post deformation, etc.)
vi) The supporting ground for the fixed

formwork to prevent geometry imperfections needs to be prepared

vii) The deck moving operation requires a great deal of technical capacity on the part of the contractor

viii) There is a high investment in the complete launching equipment which would be difficult to reuse for another bridge unless it were to have the same superstructure geometry (span length, bridge width, and depth)

ix) This methodology affects the design of the superstructure, which is governed by temporary stages during construction instead of in-service requirements

x) It requires careful geometric control during launching, mostly for lateral displacement of the top of the piers and vertical displacement of the tip of the bridge. Thermal effects are critical for launched steel bridges.

The above suggests that, for this method to be competitive, its conditions of execution need to be met rather than focusing on the economics of the materials used. Nevertheless, the economy obtained by the lack of shoring makes this method especially advisable in viaducts over deep valleys.

2.4.2.4.4 *Overall Concept and Design*

For concrete bridges, the most commonly used cross-section in works constructed by the incremental launching method is the box-girder section, as it is particularly well-suited to the shifting forces resulting from launching operations. In prestressed concrete road bridges, the ratio between the height of the deck and the span is between 1/12 and 1/16, reaching values of 1/20 using provisional intermediate supports. These ratios make the use of one launching nose whose length is usually 50 to 60 % of the span possible.

The stiffness of the launching nose influences the stress distribution in the structure and should be considered in its analysis and design.

The web thickness is usually greater than 40 cm in order to withstand the shear forces in the launching operations. The webs should be thicker in the vicinity of the final supports and the thickness of the flanges is governed by the concrete cover of the prestressing cables.

Ribbed slabs with two or more girder beams are sometimes used but, although easier to apply, are not as adaptable to shifting forces, which means that the beams are very thick, ranging from 0.50 to 1.20 metres (depending on the bridge span) to compensate for the absence of the lower flange.

During sliding operations, all sections of the deck are subjected to important changes in the sign and intensity of bending moments, as the same section can belong to both the span and the support, which implies the use of significant temporary prestress over the entire deck.

An important factor to consider is the horizontal action arising from friction during launch operations. In the case of temporary polytetrafluoroethylene (PTFE) sliding supports, the friction coefficient used in the calculations varies between 4 and 7 %. These horizontal forces only occur during the launching operations and mainly impact the design of the columns and the abutment from which the deck is launched. The resulting effects imposed on the columns, particularly the bending moments at the base, can be reduced if the columns are stabilised with temporary stay cables during the construction phase or if the sliding devices are placed with a certain eccentricity with respect to the column axis in the opposite direction to which the movement is being processed.

2.4.2.5 Precast Span-by-Span (Precast Beams or Precast Decks) with Crane or Launching Gantries

Bridges with a precast concrete superstructure (Fig. 2.29) are very popular and known practically all over the world. Nevertheless, it is interesting to observe how different countries often use diverse solutions that, in some cases, are radically different [2].

2.4.2.5.1 Field of Application

Precast beams can be economically used for spans up to 40 metres provided that the number of beams to be used is sufficient, commonly 40 units or more. This rule refers to bespoke precast beams developed especially for a given bridge and not for the application of standard beams made from a remote factory with suitable access routes.

For spans greater than 30 metres or for viaducts with piers taller than 30 metres, the concreting, conveying, and beam placing equipment must be much more sophisticated and therefore much more expensive, and is generally only used for long viaducts.

Precast concrete beams can be used either in straight or curved alignments. In the latter case, they are placed forming chords of circumferential arcs, where the concrete deck overhangs and barriers provide smooth curve geometry.

2.4.2.5.2 Advantages and Disadvantages

The main advantages of prefabrication are the ease of execution and increased quality control. Its use can lead to substantial reductions in construction times, reaching superstructure execution rates of over 3000 square metres per month.

Rapid construction with the elimination of falsework can result in a significant reduction in construction costs. However, prefabrication requires particularly careful execution, inspection, and transportation by qualified personnel, which is not always easily available.

On the other hand, prefabricated beam superstructures are generally characterised by decks consisting of numerous small cross-sectional elements with large exposed surfaces and many joints, including concreting joints, which have, in general, brought myriad durability and conservation problems to structures of this type.

In principle, the beams should all be of equal length as this facilitates production and makes it more economical. However, provided this is justified, i.e., if a sufficient number of beams is to be prefabricated, they can have different lengths, corresponding to the development of each of the alignments, which leads to some cost and labour increases although these are not usually unbearable.

Precast beams are generally of greater depth than equivalent cast-in-situ decks, either for reasons related to the construction process or because they are intended for simply supported spans, implying no slender decks. On the other hand, the generally very short side cantilevers of the cross section make the appearance of the decks even more unpleasant by accentuating this lack of slimness. In the case of curved alignments, as the precast beams are straight, these drawbacks are intensified as the side cantilevers of the decks have variable widths. [6]

Fig. 2.29 Precast beam placement with mobile crane – operations and facilities (Credits to Alvez Ribeiro and J. Lello)

2.4.2.5.3 Cross Section

The cross-section of precast beams is defined by the fabrication process, prestressing system, beam erection and placement process, and by the deck slab construction system.

For moderate span lengths, precast concrete box girders, 'π', or 'T' section girders are commonly used. They have the advantage of not requiring a cast-in-place concrete deck poured on top. The transverse continuity is provided by post-tensioning or simply grouting the joints.

However, for larger spans, these beams would become heavier and therefore the transport and lifting costs would increase. Also, the differences in deformation between the beams could make it impossible to place any transversal prestress cables.

Thus, in-situ concreting of the upper deck slab allows for lighter and therefore more easily manageable beams and is much less sensitive to different beam deformations.

2.4.2.5.4 Construction Staging Analysis

The construction sequence has considerable influence on the distribution of internal forces in the structure and therefore on its time-dependent behaviour.

In effect, the initial system is isostatic and either remains isostatic or hyperstatic after proper bonding.

The analysis of the construction phase should take into account that the superstructure is constructed in the following sequence:

1. Erection of precast beams
2. Placing of ordinary reinforcement and continuity prestressing reinforcement, if any;
3. Concreting of crossbeams and deck slab
4. Stressing the continuity prestressing cables or additional entire-span cables, if any
5. Checking the lateral stability of precast beams during installation by the designer and specifying appropriate lifting points.

It should be noted that the weight of the beams and slab is in principle resisted by the beams as simply supported, where the bending moments due to their own weight are positive (sagging).

Superimposed dead loads and live loads are applied to the concrete girder plus the concrete deck (composite section).

Concrete diaphragms and longitudinal reinforcement and/or post-tensioning (PT) are installed over the piers to make the superstructure continuous and reduce the sagging moments.

2.4.2.5.5 *Continuity Methods*

Bridge design loads can be categorised into three major components: girder weights, deck slab weights, and superimposed dead and live loads. Each of these components constitutes about one-third of the total loads, for spans in the short to medium range encountered in stringer-type bridge superstructures.

The simple span system can be structurally optimised by increasing the degree of continuity, leading to longer spans or wider spacing for the same girder depth.

Increasing the girder span or spacing for the same depth greatly reduces bridge cost. In addition, precast prestressed concrete girders can be used in long bridges.

Continuity due to girder weight can be achieved through post-tensioning using a precast pier segment.

Continuity due to deck slab weight and live load can be achieved by using threaded rod continuity methods or post-tensioning.

Continuity only due to live loads can be achieved by using reinforced bars in the deck slab over the piers. [2]

2.4.2.5.6 *Assembly*

Precast beams can be placed in several ways:

- using mobile cranes
- using launching gantries
- by lifting from the columns.

2.4.2.5.7 *Span-by-Span Erection with Launching Gantry*

Launching gantries are usually employed where the use of cranes is not possible. [3]

Generally, the delivery of precast beams to the erection front is the determining factor influencing the method of the erection system to be adopted. Launching girders, beam launchers, and lifting frames, able to receive precast beams directly behind, below, or in parallel to the erection system have been extensively utilised to erect a variety of I beams, U beams, and T beams for bridge decks. [4]

2.4.2.6 Precast Segmental Decks with Erection Gantry

Erection gantries can be used to install precast concrete segments (Fig. 2.30) span-by-span. This type of gantry supports the assembly of precast elements or segments and must support, as in other cases, the weight

Fig. 2.30 Facilities, operations and storage of precast segments

of an entire span. Transport and placement of the elements can be carried out in different ways, both independent or not of the launching gantry and whether this does or does not incorporate lifting and handling devices for the precast elements. This type of erection gantry is usually overhead, but, in special cases, it may be underslung. Depending on their position being above or below the bridge deck, the advantages and disadvantages of using these gantries are similar to those mentioned in the previous section about movable scaffolding systems, but in this case, there is no need for formwork.

The feeding system can be from behind the gantry or, alternatively, the segments can be placed on the ground underneath the gantry.

Decks are built span by span by stressing all the segments of a span together with longitudinal prestressing cables, which run the full span length. As such erection gantries need to be able to support the weight of an entire span, they are among the heaviest erection equipment used for bridge construction. [3]

2.4.2.6.1 *Field of Application*

Precast segmental bridge decks built span by span are used for spans from 20 to 60 metres, deck widths from 5 to 24 metres, and span weights from 2000 to more than 20000 kilonewtons (kN). The typical range of application is between 30 and 50 metres, with span weights between 2000 and 10000 kN and segment weights between 300 and 1500 kN. [3].

Two other issues are worth mentioning for this type of system:

- Wind during construction applied to the structure and the gantry can govern the foundation design (mostly in high wind areas).

- Special attention should be taken when determining the density of the concrete to estimate the weight of the precast segments, which might increase from the initial analysis. For span lengths in the upper range (50 to 60 metres), it could require a gantry retrofit to increase its lifting capacity.

2.4.3 Steel and Composite Decks

The most common construction methods for steel and composite decks (which, in very general terms, share some similarities with precast concrete deck methodologies) include span-by-span construction—using cranes and/or temporary supports—incremental launching, balanced cantilever, and lateral side construction. Various combinations of heavy lifting solutions can be applied to steel deck erection, as presented in Sub-chapter 5.4.

Some composite bridge solutions also involve a second construction stage, which may include, for example, the construction of concrete slabs (sometimes with lateral 'wings'). This latter approach usually requires movable scaffolding with properties similar to those used in concrete deck construction.

When building steel and composite bridges, the erection and assembly methodology to be used is one of the most important aspects to be thoroughly developed from its overall concept to the detailed design. [5]

The choice of construction methodology, in addition to economic aspects, should take into account:

- safety in construction
- the mitigation of any environmental, economic, and social impacts on the communities near the site
- the site constraints
- the manufacturing sequence and transportation of the structures
- the impact of the construction sequence on the structure's behaviour during assembly and its effects on the final structural system
- the time schedule available for the execution of the work.

Many of the aspects described in the previous section related to the construction methods for concrete decks are also valid for steel and composite decks and do not require repeating here.

The aim of the following items is to highlight the information that is specific to this type of bridge deck.

2.4.3.1 Span-by-Span with a Crane

Typically, a mobile crane (Fig. 2.31 and 2.32) is used when the profile elevation of the bridge deck to be constructed is less than 30 metres and there is a sufficient ground area for parking the lifting equipment within a radius of less than 25 metres between the equipment and the final position of the element to be assembled.

At the same time, vehicle access must be guaranteed to all areas where crane parking is foreseen.

As the crane's capacity is greatly reduced when lift height and range are increased, more careful consideration of the feasibility of this method is required for larger heights and ranges.

The most common phases, when opting for this methodology, are the pre-assembly on the ground of the deck elements to be raised, their elevation and positioning, and, if applicable, their connection to the previously erected and assembled elements.

Fig. 2.31 Pre-assembly of the steel structure

Fig. 2.32 Erection of steel girder with mobile crane

Shoring or other types of temporary ground support are often required to ensure the geometry and structural integrity of the structure to be assembled, as well as safety during the construction process.

2.4.3.2 Segmental Decks with Temporary Supports (Static Falsework)

As previously covered in the precast span-by-span concrete decks chapter, this bridge typology and the associated lifting equipment are standard. Moreover, for steel bridges, the splice connections are quite common for the installation of the steel girders. This allows for the installation of long girders which could otherwise be challenging to put in place in one piece. A conventional system is to install, on a temporary structure, the superstructure pier segments over the piers, which are usually much heavier than the mid-span segments. Later, a drop-in segment can be installed in the middle.

Temporary towers are usually located at points of low bending moment in service (L/4 or L/5), where the bolted connection design will be optimal. However, it will also change the design of those sections during construction, now required with hogging moments due to the temporary support. This temporary condition also affects the splice connections during the drop-ins, as the girders will be fabricated with the fabrication camber and connected in the air with a different rotation.

The span length for steel bridges erected with cranes on temporary supports could go up to 90 m.

2.4.3.3 Incremental Launching

The incremental launching methodology (Fig. 2.33 and 2.34) is especially competitive for large viaducts with a regular layout (flat geometry) where the profile elevation of the bridge is high and where there are access restrictions or environmental constraints.

The construction rate associated with this methodology primarily depends on the number of main girder elements to be welded on site, as the transversal beams or diaphragms and bracing elements are usually bolted. Some countries where bolt connections are quite common (like the United States of America or Canada)

Fig. 2.33 Structure push system

have developed systems to launch bolted girders, despite the bolt connections at the bottom flange.

For spans between 30 and 55 metres, involving the execution of two or three welded joints per girder, one span can be launched every 9 to 12 working days.

Thermal effects challenge geometric control due to vertical and lateral displacements of the tip. At piers, lateral restraints should be installed.

As an alternative, king posts can be used for steel bridges instead of the launching nose. In some cases (as in the Millau Viaduct) the permanent pier was used as a king post.

2.4.3.4 Balanced Cantilever Method: Lifting Deck Segments

In steel and composite bridge deck construction crossing rivers or the sea, one of the options is to bring successive deck segments by barge and lift them into position (Fig. 2.35).

This type of balanced cantilever method takes advantage of heavy lifting technology. Cantilever bridge segment lifter equipment usually provides a safe, economical, and efficient way to lift and hold bridge segments in place over water when the use of a crane is not feasible. The equipment is installed on the bridge deck and lifts the steel girder section from the transport barge (or from a trestle bridge) to the bridge deck for assembly and welding. The system cantilevers off the last segment of the bridge that was constructed over land. Once the new segment is in place over the waterway, the cantilever segment lift equipment is then relocated to the newly installed bridge section so that the next bridge segment can be lifted into position.

Fig. 2.34 Front end of frame during launching

The process is conceptually simple but the dynamics of the operation may be rather complex, particularly in open waters, as in a bay or estuary. During lifting, both the barge and the bridge undergo mo-

Fig. 2.35 Cantilever segmental bridge lift system (photo: https://www.huadacrane.com)

Fig. 2.36 Lateral slide of bridge deck into its final position

tions induced by sea currents and waves and by wind action. A successful outcome requires limiting the misalignment of supports, the plinth loads, the uplift loads in the erection traveller, the dynamic load factor, and the forces developed in the hoisting and retaining cables.

In some cases, the wind was seen to play a relatively unimportant role compared with the waves. A critical phase is when the segment is about to lose, or has just lost, contact with the barge, so this phase should be kept as brief as reasonably possible.

Segment lifting equipment is usually designed to install segments that weigh less than 250 tonnes, over waterways and in varied design configurations in tight locations, or in areas that are generally difficult to access.

Cantilever segmental bridge lift systems are more versatile than a launching gantry. They are easier to construct and lighter because it is not required to span from one pier to the next.

2.4.3.5 Lateral Slide

Although relatively rare, the lateral slide process (Fig. 2.36) is particularly suitable when replacing existing decks because it has several advantages over all other alternatives.

The main advantage is that it allows the existing bridge to be kept up and running without traffic limitations during construction and to be replaced in a short time, mitigating the impact of the work on traffic.

2.4.3.6 General Comments

A series of general remarks are worth mentioning for steel and composite deck bridges:

- Fabrication cambers are highly dependent on the means and methods for the erection of the superstructure
- Wind during construction usually requires designing temporary bracing. This bracing restrains the girders and can be removed as soon as the concrete deck is poured
- To ensure lateral stability, long girders lifted with the span-by-span method should be installed in pairs for the first two girders, or with a temporary structure on top to restrain the top flange

- Steel box girders are commonly used for curved alignments to accommodate torsional effects
- For long-span girders, orthotropic decks are required instead of concrete decks, to minimise weight and optimise the design.

2.4.4 Lessons Learnt and Recommendations

As detailed in other chapters, including Section 4.2.7 on bridge-equipment interaction and Section 3.1.7 relative to geometric control (especially for cantilever methods), the following points should be addressed:

- As a general note, it is crucial to foster close integration between bridge design and construction equipment design. This coordination can significantly save time and costs while reducing operational risks. It is necessary to ascertain the construction stage loading and reactions from equipment to determine the structural adequacy and stability of the partially completed structure.
- Comprehensive risk analysis is essential in bridge projects near aerial electric infrastructures. This analysis should extend beyond the standard clearance requirements for bridge elements and construction equipment. It is vital to consider the kinematics of lifting equipment, like cranes. This is a high-incidence cause of construction accidents
- For bridge decks constructed with ground scaffolding, attention must be paid to the potential lateral sliding of the scaffolding's base. This issue can arise from changes in ground soil characteristics due to rain or from severe transversal forces, often associated with water lines during floods
- In the case of bridge decks erected using incremental launching on a descending slope, careful consideration of the braking system is required. It is important to account for dynamic amplification during braking and monitor friction levels
- For bridges using movable scaffolding systems or launching gantries, the considerations mentioned above should also be applied to the equipment design and reviewed by the bridge designer
- When constructing bridge decks with precast segments using short-line precast, special attention must be paid to consistent geometric defects in the moulds. Even if these defects are within regulatory limits, they can lead to geometric incompatibilities
- In the case of in-situ, span-by-span construction of prestressed concrete decks using deformable equipment (L>500), the occurrence of cracks in the upper fibres at mid-span during the prestressing application stage must be addressed
- It is necessary to monitor bridge deck defects during each stage of construction, particularly in cantilever construction
- In composite bridges, although not all codes prohibit onsite welding operations, it is generally advisable to avoid them unless stringent quality assurance procedures are in place.

References

[1] The Constructor. Balanced Cantilever Method of Bridge Construction – https://theconstructor.org/structures/balanced-cantilever-method-bridge-construction/6137/
[2] Conceptual Design of Precast Concrete Bridge Superstructures, fib (fédération internationale du béton) Technical Report, Task Group 6.5
[3] Bridge Deck Erection Equipment, A best practice guide. IABSE Working Group 6, ICE Publishing
[4] Bridge Construction Partner, VSL International Ltd
[5] Rui Alves (2019). Metodologias de montagem de obras de arte metálicas e mistas
[6] Cristina Machado. Processos Construtivos de Pontes de Médio Vão, I.S.T., U.T.L.
[7] Miguel Ferraz. Um Modelo de Análise para o Estudo de Pontes como Estruturas Evolutivas, F.E.U.P.
[8] Pacheco, P., (2015). "Multi-span Large Decks – the Organic Prestressing Impact", Proc. Multi-Span Large Bridges, CRC Press, Porto, 103–124
[9] Engineered Rigging. Cantilever segmental bridge lift – https://engineeredrigging.com/equipment/cantilever-segmental-bridge-lift/
[10] Joaquín Martí (2018). Lifting deck segments for bridge construction over water – Principia – https://principia.es/en/lifting-deck-segments-for-bridge-construction-over-water/
[11] AASHTO. Construction Handbook for Bridge Temporary Works, 2nd ed. 2017
[12] European Standard EN 1991-1-6:2005. Eurocode 1: Actions on structures. Part 1–6: General actions – Actions during construction.
[13] European Standard EN 1991-1-7:2006. Eurocode 1: Actions on structures. Part 1–7: General actions – Accidental actions.
[14] Rosignoli, M. (2013) "Bridge Construction Equipment", ICE Publishing
[15] Lucko G., Garca J. M. D. L. (2003) "Constructability Considerations for Balanced Cantilever Construction", ASCE, J. Structural Design and Construction, Vol.8, Issue 1.
[16] "The cantilever construction of Prestressed Concrete Bridges" Jacques Mathivat
[17] Podolny and Muller "Construction & design of prestressed concrete segmental bridges"
[18] EN 1337-11:1997 Structural bearings – Part 11: Transport, storage and installation

2.5 Modular bridges

Inês Ferraz[1] | John Kerkhoff[2]
[1] Portugal, [2] Netherlands

2.5.1 Introduction

The majority of bridges constructed worldwide are not part of major infrastructure networks or spectacular works of art. Instead, most bridges consist of minor spans and narrow lanes. Despite their modest nature, these bridges are vital for the lives of millions of people, serving as the physical link between communities. They permit the inflow and outflow of people and goods, resulting in a significant socioeconomic impact. They also serve as infrastructure support for cables and pipes, and for special transport like that used for defence purposes and railway traffic. They can even constitute temporary movable bridges when shipping is an important factor.

For all the above purposes, modular bridges are many times the chosen option as it is a prefabricated, standardized bridge system composed by interchangeable components that can be quick and easily assembled and disassembled, both for temporary and permanent use and are transportable in standard containers. They can be made of steel, aluminium, or composite materials, offering versatility in span length and load capacity. Somehow, modular bridges can be seen as a particular case of DfMA bridges which refer to bridges designed and constructed using the "Design for Manufacture and Assembly" principles, which focus on simplifying the design and optimizing the manufacturing and assembly processes to enhance efficiency and reduce construction time and costs.

Since the mid-20th century, modular bridges have become a popular choice for constructing bridges quickly, easily, and economically. They have been chosen for various circumstances, such as:

- Emergency situations, for example, a collapse due to military intervention or natural disaster, where rapid restoration of infrastructure is crucial for military operations or evacuating people
- Logistical support
- Provisional detours, used while permanent bridges undergo construction or maintenance
- Temporary use, pending the construction of a permanent solution. Often, the duration of these installations extends beyond the initially planned period
- Permanent use, when designed for from the outset.

The fast construction achieved by a modular bridge allows for the quick restoration of infrastructure in all the cases described above. This is a fundamental factor in promptly meeting the need for creating new routes and logistical transport circuits.

Simultaneously, the effects of adverse weather conditions, natural disasters, and

Truss Bridge MB60 by MBSbyBERD, Peru (photo: MBSbyBERD)

Panel Bridge System by Mabey, Portugal (photo: MBSbyBERD)

Beam Bridge System, Netherlands (photo: Janson Bridging)

◀ **Fig. 2.37** Examples of modular bridge types [1–5]

conflicts are increasingly generating the need to create alternatives that maintain the free movement of people and goods.

In this context, modular steel bridges are, and may increasingly be, the solution to efficiently solve these 21st-century problems. Their effectiveness lies not only in their fast and easy assembly but also in their alignment with the current trend of modularisation in the construction field, which is becoming a new-fangled construction method. Besides, one of the most distinctive features of modular bridges is that they are designed for reuse, contributing to a circular and environmentally friendly material management system. Fig. 2.37 provides some examples of existing solutions.

The modular bridge is a general concept that may be materialised by several structural systems and may be constituted by a wide variety of pieces/geometries. For example, they can be panel type, truss type, or beam type, provided they are composed of light and transportable pieces and easily assembled and disassembled.

The following sections introduce and discuss the main issues of design, assembly, and installation and present an analysis of modularity for efficient management.

The main references for this section are [1–5].

2.5.2 Historical Perspective

Up to now, modular bridges have often been associated with and confused with the Bailey bridge system, despite not being the same. A modular bridge can take on various geometries and applications as long

Fig. 2.38 Donald Bailey (left) and Bailey Bridge Assembly in 2nd World War (right)

as it is composed of elementary pieces or modules that are industrially produced in a factory, transportable to the site, assembled with light equipment employing basic manpower, and is possible to be dismantled for reuse at another location. However, the Bailey bridge system was indeed the first successful concept of a modular bridge, and it remains applicable today with slight modifications since its inception.

During the Second World War, Sir Donald Bailey was tasked with developing a new bridge system that could meet the increased demand for bridges. This demand was spurred not only by extensive bombardments that damaged numerous bridges but also by the introduction of heavier tanks used by the Allies.

These bridges needed to have lightweight elements, be assembled by hand only, be easily transported, follow simple fabrication standards, and avoid using aluminium, as this was a critical resource for aircraft manufacturing.

Thanks to the limited number of different elements, the concept devised by Sir Bailey (Fig. 2.38) also facilitated the storage of materials that could be rapidly deployed in emergency situations, accommodating various spans and widths.

The concept was extensively used during the war and has been maintained until today due to its undeniable flexibility, as these bridges can be easily reinforced and extended in length by adding more panels to the bridge configuration.

However, in recent years, other bridge concepts have been successfully developed, challenging the long-standing 'monopoly' of Bailey bridges.

2.5.3 Modular Bridges – Design and Lifecycle Essentials

2.5.3.1 Standard/ Dimensions

Bridge dimensions are obviously adapted to customer requirements, both for public and private entities, following the rules of local standards. Nevertheless, in the case of Bailey bridges, designed and developed in England during World War II, their geometry was originally specified in feet (ft), and usually respects the following standard dimensions, presented in Table 2.2 in metres (m). Till today these standard dimensions remain usually applicable.

Table 2.2 Typical dimensions in most common Modular Bridges

Lane Number	Width (m)	Span
1 (single lane)	3.15	Up to 60.96 m (=200 ft.), multiples of 3.048 m (=10 ft.)
1 (single lane, extra-wide)	4.20	
2 (double lane)	7.35	

2.5.3.2 Structural Systems and Elements

As noted above, several structural systems can be used to qualify as modular bridges. Panel bridges typically employ a half-through design, where the bridge is composed of two longitudinal trusses that support transverse beams on their lower chords, thereby supporting the deck panels. These longitudinal trusses can be made up of panels of various geometries or any other prefabricated elements aligning with the principles of modularity. Longitudinal and transverse bracing, provided by secondary structural elements, ensures both local and overall stability of the bridge.

Beam or girder bridges are composed of steel beams located between the two abutments, with the steel deck modules being assembled over them.

Truss bridges can be through-type or half-through. The former consists of a box truss girder and the latter entails two longitudinal lateral trusses. In both examples, the vehicle flow occurs at the lower level of the section.

2.5.3.3 Module Size and Weight and Logistic Restrictions

Modular bridges are usually designed to fit within standard 20-foot or 40-foot ISO containers. Since the internal dimensions of these containers are approximately 2.40 metres in height and 2.35 metres in width, all structural elements must conform to these sizes.

The weight of individual elements is also a consideration. Many bridge suppliers maintain the idea that a bridge should be assembled using only manual labour, like the original Bailey bridge, thereby limiting the weight to approximately 400 kilograms per element. However, with the increasing use of cranes and other lifting equipment as well as the requirement for safe working conditions, the significance of this weight limitation has diminished. The 20th-century practice of weight limitation is becoming less applicable in situations where suitable mechanical handling and lifting means are available, which is increasingly common in the 21st century. However, the use of small lifting equipment and limited use of manpower is still seen as an advantage.

For long-distance transportation, the most commonly used means are ISO containers. Trucks are used for shorter distances, including the delivery to the final location. Elements can be supplied in separate parts or in pre-assembled sections, depending on local conditions. The load of the containers is often maximised during the design stage to reduce logistic costs, which have been rising and unstable since the Covid pandemic.

2.5.3.4 Load Capacity and Design Codes

The load capacity of a bridge must be defined considering its use—military or civil.

For civil applications, civil loads and codes must be applied. Worldwide, it is not rare that the standard set by the American Association of State Highway and Transportation Officials (AASHTO) is accepted, even outside the United States of

America. However, most countries enforce their own standards supplementarily. This is equally the case in Europe where the Eurocode must be followed, Canada (Canadian CAN/SA-S6), Australia (AS5100), or South Africa (SATCC). In the case of temporary applications, adjustments or reductions to the standards may be applied, when properly framed within legislation.

For military use, the applicable standard applied is STANAG2021 (NATO standard), which defines the military load classes (MLC) for both tracked and wheeled vehicles. The trilateral design and test code for military bridging and gap-crossing equipment, approved by Germany, the United Kingdom, and the United States of America, may also be applicable.

Permanent concrete abutment (photo: MBSbyBERD)

Alternative option for non-permanent bridge (photo: Janson Bridging)

Fig. 2.39 Abutment examples

2.5.3.5 Banks and Abutments

For permanent modular bridges, abutments are typically designed in classical terms. In the case of non-permanent modular bridges, solutions delivered by the supplier sometimes include the actual banks (Fig. 2.39).

In both cases, the banks must have the necessary bearing capacity to support the bridge. Although these structures are lighter than typical concrete bridges, the pressure on the soil should be under control. Visual and simple tests can be conducted to estimate bearing capacity. The removability and reusability of the temporary abutment and intermediate supports are often considered when making the design and choice of materials.

2.5.3.6 Supports

Final or permanent supports can also have different solutions. Traditionally, for panel bridges, bearings are usually metallic, pinned or hinged, allowing the bridge to bend longitudinally, which defines the bridge as simply supported. However, and mainly when alternative bridge solutions are considered, other types of bearings are applied, such as elastomeric or customised, which permits adaptation to local requirements, for example, related to seismic loads (Fig. 2.40).

There are also provisional supports used only during launching (discussed further below), which are not part of the modular bridge itself, but part of the assembly kit.

Metallic hinged bearing
(photo: MBSbyBERD)

Elastomeric bearing
(photo: Janson Bridging)

Fig. 2.40 Examples of Modular Bridges Permanent supports

2.5.3.7 Manufacturing Geometric Control

For the successful implementation of modular bridges, it is essential that appropriate geometric control is implemented during all phases of the manufacturing process, from cutting, drilling, welding, and after galvanisation, as all these steps have an impact on the dimensions of elements.

The importance of geometric control is not only related to the fitting of all the elements together during assembly, as all clearances are very tight but also refers to the bridge's behaviour while being launched. Extra care should be given to the tolerances in chords as they slide over wheels or rollers, and the supplier and installer should have a sound experience in structures that are incrementally launched.

2.5.3.8 Modularity Degree

Modularity is a fundamental attribute of modular bridges, so intrinsic that it is reflected in their name. However, modularity should not be considered an unquestionable conceptual value; instead, it must be correctly and rationally evaluated on a case-by-case basis.

The modularity degree can be expressed by Equation 1:

Eq1

$$MD = \frac{Total\ Number\ of\ Pieces\ of\ a\ set}{Total\ number\ of\ Different\ pieces\ of\ the\ same\ set}$$

It is known that if the modularity degree (MD) is high, then:

- in any future applications, the costs tend to decrease
- if any future applications are required for a set of bridges with different requirements, the costs also tend to decrease.

However, it is also true that if the modularity degree (MD) is high, design optimisation tends to be lower due to the use of many oversized parts, and then:

- the initial costs are (not possibly, but in fact) higher
- the initial CO_2 emissions are (not possibly, but in fact) higher.

The MD value for modular bridges may vary according to the span, width, and concept. Usually, a minimum value is greater than 10 and can reach around 60. Another important indicator is the number of average assemblies per bridge (NA) in a population (or set of modular bridges) in a reference period, for example, 5 years, as expressed by Equation 2.

Eq2 $$NA,5 = \frac{\textit{Total Number number of Registered Assemblies of the Bridges in a Set}}{\textit{Total number of Bridges in a Set}}$$

If, among a population of modular bridges, an NA,5 is expected to be NA,5 > 3, higher MD values are normally required (possibly MD>20). Conversely, if an NA,5 is expected to be NA,5 < 1.5 among a population of modular bridges, lower MD values are typically required, for example, MD >5.

For higher values of MD, it is important to perform a critical analysis, as the increased modularity may result in additional costs, both economic and in terms of CO_2 emissions. Therefore, if the use of the bridge during its life cycle is fully defined at the initial stage, it is advisable to optimise the structure to suit its intended use and reduce the degree of modularity.

2.5.3.9 Permanent Versus Non-Permanent Bridge Specifications

Modular bridges are often seen as temporary solutions, possibly due to their lightness and easy disassembly. However, this is a misconception as the concept of modularity does not refer to the expected period of use. Modular bridges can be both temporary and permanent structures.

The definition of a bridge as provisional, semi-permanent, or permanent varies among standards around the world, where the design life for a permanent structure can be from 50 to 75 years (or more for special bridges) and may range from 5 to 25 years for a provisional bridge. In the case of modular bridges, durability is simply the result of specifications, meaning it depends on the bridge owner to specify bridges for permanent use.

In general, the classification of the structure occurs during the design phase and involves considerations such as seismic and wind impacts according to recurrence periods, fatigue issues like the number of cycles, collision impact, and factors related to redundancy and importance of the bridge. All these considerations typically affect the cost of the bridge.

It is crucial to note that provisional bridges must be designed to the same structural safety requirements as permanent bridges.

Maintenance is also a key issue for life expectancy. Suppliers should provide an indicative maintenance programme that must be followed and adapted by the bridge owner. It involves the definition of inspections and their periodicity as well as the follow-up actions to be taken, like repair, retightening, cleaning, and replacement. In fact, one of the advantages of modular bridge systems is being built with standard items that, in the event of damage due to wear or collision, can be replaced relatively easily and quickly from available stock.

The classification of a bridge as permanent or non-permanent should be clearly determined by the client during procurement to ensure appropriate measures in terms of design and maintenance. Lack of a clear definition should be avoided, as it can lead to risky situations. A common scenario is a modular bridge installed as provisional with a specific time frame being used far beyond the initially defined number of years.

2.5.3.10 Exposure to Floods and Water Lines

When studying the implementation of modular bridges over rivers or other water bodies potentially subject to floods, conducting a risk analysis is crucial. Typically, in such cases, the following decision paths should be considered:

- Ensuring an appropriate topographic level of the bridge, possibly with landfill in access areas
- Analysing the span length or multi-span solutions, considering the insertion of piers
- Evaluating the risk acceptance of a full collapse, including its consequences and costs
- Assessing the risk of extremely high water
- Assessing the socioeconomic impact of a go/no-go decision
- Consider the use of the river by ships and possible impacts.

Numerous problems may arise if these issues are not properly addressed.

For instance, situations where bridge-span limitations lead to an insufficient length and to abutment erosion by water flow, are common. This explains the current trend of increasing span lengths whenever permitted by economic conditions.

Also, if the level of the bridge is not correctly defined, accidents may occur with water flow displacing the entire bridge, or due to the impact of floating objects (driftwood, ice, ships, etc). Hydrologic studies are essential to determine the bridge height, as modular bridges are typically not designed to withstand the force of water currents. It is crucial for clients or designers to ensure a low probability of water levels reaching the bridge, especially in rivers with unstable flow.

Another example of the impact of these decisions is: If the height of access roads to the bridge, within a reasonable distance (for example, within a 25 km radius from the bridge), is below the flood level, it may not make sense to elevate the modular bridge and surrounding access roads.

2.5.3.11 Sustainability and CO_2 Emissions

All players should consider sustainability values, especially regarding CO_2 emissions, embodied in production energy, material consumption, waste, reuse, and recycling of the material used. The promoting entities have the most significant impact and responsibility in this process.

It has been shown in publications that when an optimised solution is adopted instead of a standard Bailey concept, an average reduction of over 8 % in CO_2 emissions is achievable through such duly evaluated options [2], but this value can be greater than 20 % in many instances. Therefore, the most eco-friendly structure is probably not the one with the highest degree of modularisation.

2.5.4 Installation/assembly of modular bridges

2.5.4.1 Construction Methods

The installation of main girders for modular bridges may be performed according to the following classical bridge construction methods, with very different frequencies:

- A → Incremental launching – most common for panel bridges and truss bridges
- B → Full segment lifting or crane mounting/installation – standard for beam bridges and not rarely for panel and truss bridges
- C → Partial lifting – hoisting of main girders one by one and assembly transoms and deck afterwards – standard for beam bridges
- D → Full segment basculation – very rare
- E → Balanced cantilever – very rare
- F → Hybrid combinations of different methods and others.
  Example 1: Incremental launching assisted by crane – not rare;
  Example 2: Float-in with a floating crane or pontoon – rare

In general, these methods all comprise: 1) preparation of the assembly yard; 2) preparation of banks; 3) installation of supports; 4) assembly of the main girder in situ; 5) final positioning of the main girder; and 6) assembly of deck modules and other secondary elements. In the following points, different steps are discussed exclusively for the more common methods A and B.

2.5.4.2 Assembly Yard

The installation yard should be cleaned and levelled sufficiently to receive containers or trucks with parts and also to facilitate the assembly and operations of the chosen construction method (for example, the launching operations). In the case of incremental launching, following the bridge guideline, an assembling corridor with a minimum size equal to the bridge length must be ensured on the launching bank. If the space is restricted, the length of the assembly platform can be reduced by performing a higher number of partial assemblies / partial launchings and using a higher counterweight. On the opposite bank, the yard should provide the same features and allow the disassembly of the launching nose. The width of the yard should be sufficient to accommodate the bridge assembly, including lifting equipment operations, and should never be less than double the bridge width, as a general guideline.

2.5.4.3 Banks

As mentioned above, banks must possess the necessary bearing capacity to support the bridge not only during its utilisation phase but also throughout the installation phase. Despite being a light structure, the pressure exerted on the soil must always be controlled. For more details, see Section 2.5.3.5 Banks and Abutments.

2.5.4.4 Incremental Launching

2.5.4.4.1 In-situ Assembly

On one of the banks, the bridge is assembled (Fig. 2.41) and then launched until it reaches its final position. The process

Fig. 2.41 Main trusses ex-situ assembly

Panel Bridge pinned connection
(photo: Janson Bridging)

Truss Bridge bolted connection
(photo: MBSbyBERD)

Fig. 2.42 Assembly – Connections

starts with the installation of temporary supports (usually wheels or rollers), followed by the assembly of the launching nose and the main structure of the bridge. This structure includes trusses, transoms, and longitudinal and transverse bracing. The incremental launching process, balanced with counterweights, is used to position the bridge at its final longitudinal location. Following this, it is necessary to lower the bridge from the level of the wheels to the level of the final supports. Only at this point the rest of the structure like deck modules or add-ons like walkways and rails is assembled.

Connections between structural elements are primarily pinned or bolted (Fig. 2.42). Welded connections are permitted in the workshop but are never allowed on the construction site.

Launching process

Incremental launching is the more common method for installing panel and truss bridges (Fig. 2.43).

The launching process is based on a simply supported structure with a cantilever balanced by a counterweight installed in the assembled rear part of the bridge, as per the conventional 'incremental launching method.'

Special care must be taken regarding the nose 'landing' stage, where suitable studies and know-how ensure proper matching, both at the topographic level and plan view.

In most installations, the launching process is facilitated by basic machinery available at the construction site, as shown in the image above, where multifunctional lifting equipment possesses the capacity to push the bridge.

2.5.4.4.1.1 Launching Nose

The launching nose (Fig. 2.44, right) is a lightweight temporary structure used only during the launching process to permit the bridge to reach the opposite bank without introducing high internal forces that would compromise the structure. The deflection of the nose is not an issue, provided counter deflection is accounted for, through the ski geometry.

The launching nose may or may not be constructed with the same items as the bridge, which underlines the modular nature of the system

2.5.4.4.1.2 Crane Assistance

During the assembly and launching process, the assistance of a crane (Fig. 2.44,

Launching, lateral view (photo: MBSbyBERD)

Launching, back side (photo: MBSbyBERD)

Fig. 2.43 Incremental Launching operation

Launching nose
(photo: MBSbyBERD)

Crane Assistance, module pre assembly
(photo: MBSbyBERD)

Fig. 2.44 Launching nose and crane operation

left) greatly accelerates bridge installation and increases working conditions and safety, although manual assembly may be possible in many cases. Cranes or multifunction lifting equipment are typically used during the assembly and dismantling operation. The lifting capacity must be determined by considering the maximum weight of the elements and the operational distance.

2.5.4.4.1.3 *Temporary Supports*

During assembly and launching operations, the bridge rests on temporary supports equipped with a rolling (Fig. 2.45). The bridge moves or rolls in controlled steps until it reaches its final position. Only then are the temporary supports replaced by the final ones at the abutments and intermediate supports.

Temporary supports are necessary to provide a rolling line where the bridge can be pushed, typically consisting of rollers, wheels, or any other rolling system, usually applied over bogies. These elements facilitate rotation and allow the adjustment of support elevation in the launching yard to accommodate any fabrication camber, if present, during the different launching stages. When bridges are too heavy for the standard construction rollers, methods such as skidding could be considered.

Fig. 2.45 Launching supports – Bogies with wheels (photo: MBSbyBERD)

Fig. 2.46 Counterweights with deck modules (photo: MBSbyBERD)

2.5.4.4.1.4 *Counterweights*

Often, the balance of the bridge during the launching process is not ensured by the weight and length of the bridge alone, and counterweights are required (Fig. 2.46).

The deck modules are commonly used at the back of the bridge to serve this purpose. After completion of the launching, these deck modules are assembled in their final position.

2.5.4.5 Full Bridge Lifting and Positioning with Cranes

Similarly to the launching process, the method of using one crane or a pair of cranes to install a bridge in a single operation is very common, especially in places where cranes are readily available and economical. This method begins with assembling the bridge on the bank or any available area or by transporting fully pre-assembled bridge sections to the site, if possible. Then, the crane lifts the bridge at four points (Fig. 2.47), two on each truss, and through a combination of translation and rotation movements, installs the bridge in a single operation. This operation is frequently carried out with the bridge partially assembled to limit the required crane capacity. The decks and other secondary parts are only installed afterwards.

The capacity and cost of the crane are obviously major considerations, but using cranes tends to lead to a very fast process and less hindrance for the area. High-capacity cranes can install longer modular bridges; however, their size, weight, and rental cost become significant factors. Generally, in the case of panel, beam, and truss bridges and only for spans up to 30 metres, this process is an alternative to the launching method.

Lifting ties, made of steel cables or ropes, must be prepared with soft materials where they contact the structure to avoid damaging the hot-dip galvanize or other corrosion protection. The lifting points must be carefully considered during design to prevent unpredictable forces on the structure, and designed lifting beams must be used.

2.5.5 Lessons Learnt and Recommendations

The following notes outline an analysis of the entire process of designing, constructing, and assembling modular bridges. They particularly emphasize the thorough and critical examination of the needs and limitations associated with construction and assembly and how these factors influence the development of more rational solutions.

- **New rules for Tender Specification**

It is observed that many public tenders are currently written in a very strict manner, specifically for Bailey bridges. However, tender specifications should be adapted to the reality of the 21st century. They should be open to engineering optimisation by being based on performance defined by maximum vehicle weight, number of axles, fatigue, deflection requirements, etc., rather than being based on a specific concept or product. This approach will lead to the delivery of bridges that are structurally more efficient, more sustainable, and adapted to client needs.

Furthermore, in the context of modular bridge specifications, the practice of imposing a weight limitation per piece is not relevant in situations where appropriate mechanical handling and lifting means are available.

- **Suppliers' Qualification**

In the context of modular bridges, it is crucial to ensure that suppliers have relevant experience with structures that incorporate modularity. Additionally, they should possess significant experience with incrementally launched structures and uphold high-standard procedures for geometric control in steel manufacturing. A good implementation and safety plan with extensive engineering should be provided as a minimum to achieve a safe and good solution.

Full bridge crane lifting (photo: Janson Bridging)

Special lifting beam (photo: Janson Bridging)

Fig. 2.47 Crane Lifting and Lifting Beam

- **Material Quality**

Current regulations often require traceability of materials, material quality, and production quality marks. Old systems, such as Bailey bridges, cannot meet current standards.

- **Rational Degree of Modularity**

In the design of modular bridges, an excessive increase in the degree of modularity tends to raise initial costs and CO_2 emissions, while it tends to reduce costs in re-assemblies. Therefore, the degree of modularity should be adjusted according to the average number of expected assemblies per bridge.

- **Modular Bridges: An Eco-Friendly Option**

Compared to other solutions, such as concrete bridges, steel modular bridges are a better option in terms of life cycle analysis. Being modular, they allow for disassembly and reinstallation or reuse. Their metallic nature enhances the possibility of recycling. Additionally, as they are lighter, they reduce the consumption of materials in abutments, piers, and foundations.

- **Modular Bridges as Permanent Bridges**

Modular bridges can also be regarded as permanent bridges, provided they comply with standard specifications related primarily to fatigue and corrosion.

References

[1] A.André, J.Fernandes, I.Ferraz; P.Pacheco: New Modular Bridges Solutions, 9th International Symposium on Steel Bridges; IOP Conf. Series: Materials Science and Engineering 419 (2018) 012021 doi:10.1088/1757-899X/419/1/012021

[2] A.André, J.Fernandes, I.Soares; P.Pacheco: New Modular Bridges Solutions—A sustainable Solution to Connect People, DOI:10.1007/978-3-319-70272-8_44

[3] A.André, I.Ferraz, H.Coelho, J.Fernandes, P.Pacheco: New Span Range on Modular Bridges – The Arch Relevance, DOI: 10.1007/978-3-030-29227-0_103

[4] A.André, I.Ferraz, H.Coelho, J.Fernandes, P.Pacheco: The 125 Modular Bridges Project to Peru, DOI: 10.1002/cepa.1189

[5] R.Stegmaier: Maximum Load Capacity of Bailey Bridges, DOI:10.1061/TACEAT.0006589

3

Special Bridge Construction Methods

https://doi.org/10.2749/sp001.ch03.1

3.1 Cable-Stayed and Extradosed Bridges

Miguel Angel Astiz[1] | Marcos Sánchez[2]

[1] Spain, [2] Ireland

3.1.1 Introduction

Cable-supported bridges (either suspension, cable-stayed, or extradosed) share the property of exhibiting different structural systems during construction and service life. While the final structural systems are very efficient, these bridges must incorporate other systems during construction which are not as efficient, and which may be more sensible to ambient actions such as wind or earthquakes. In many cases, these systems are not very robust and require the addition of temporary elements such as temporary piers or links to provide the minimum reasonable robustness to the structure. Relevant information on cable stayed bridges is presented in [1–9].

Cable-supported bridges also share the property of being very flexible and, consequently, may undergo significant displacements during construction when there is no continuity along the whole deck and only a fraction of the cables have been installed. These displacements may cause critical temporary internal forces in the deck, and may even induce relevant geometrical non-linear effects.

Cable installation is not simple either since, during construction, cable stresses may show large variations, making it necessary to impose limitations and constraints on the different operations in addition to the limitations imposed by the deck stresses. Besides these constraints, it must be ensured that the cables are not damaged during construction to preserve their long-term resistance, and, consequently, all the steps of cable load variation should be done by avoiding "biting" (i.e. physical penetration) on the working length of the cable.

Finally, the construction of complex structures such as cable-supported bridges requires the ongoing monitoring of the main variables of the process: displacements, cable loads, deck stresses, etc. These data must be compared to design values to ascertain that the final bridge geometry and stress distribution are as assumed in the design. Cable-stayed and extradosed bridges are highly hyperstatic structures, and the final stress distribution may be quite different from the design, even if the final designed geometry has been achieved [6]. In the case of suspension bridges, especially for long spans, it is easier to achieve the designed stress distribution, but the main goal may be to obtain the designed geometry. In both cases, construction monitoring and permanent comparison between measured values and design values is a necessity.

All these aspects are being considered in this chapter. The deck erection methods have been categorized into three classes which are not all of the same importance, but are very different from a conceptual point of view: construction on fixed supports, incremental construction, and segmental construction. In the first case, the deck is completely built at the first stage and cable forces are applied afterwards. In the second case, the deck is built at one of the abutments and is pushed forward sequentially and together with the towers, involving rather complex operations. In the third case the deck is built segment by segment and cable forces are applied simultaneously. There is a fourth type of construction, which generally corresponds to long-span suspension bridges, in which the cables are erected in a first phase and the deck is erected, segment by segment, afterwards. This type of construction could also be called segmental construction although a better understanding of the conceptual differences may be reached if it is called cable-dominated construction. This last type of construction is not considered in this chapter, which concentrates on cable-stayed and extradosed bridges.

Section 3.1 is mainly focused on describing and explaining the essentials on construction methods applicable to cable-stayed and extradosed bridges. For a more in-depth analysis of this type of bridge, an extensive bibliography is available, as shown in the list of references at the end of this section [7–9].

3.1.2 Construction on Falsework (Fixed Supports)

3.1.2.1 Construction Description

In certain circumstances, the preferred option may consist of constructing the deck on fixed scaffolding. This may happen if the deck height over ground level is low and the soil is sufficiently rigid to support temporary piers or scaffolding by means of simple spread foundations. Construction costs may also govern the decisions since in some countries it may be advantageous to build scaffolding instead of using more complex methods such as the cantilever method. This method has mostly been applied to urban bridges or overpasses. In both cases, the height is low (slightly greater than the required clearance) and the span length is comparatively short; many of these bridges are just small iconic bridges.

The principle of construction is very simple. In the first phase (Fig. 3.1), the abutments, piers, towers, and deck are built: all the elements of the bridge are in place except the cables. Only the deck is supported by temporary supports. This continuous support is either full-volume scaffolding or a stiff beam on temporary piers. The decision between these two alternatives should depend on an economic optimisation in which the soil rigidity and the height of the deck may be the most important constraints. In the second phase, the cables are installed and stressed (this operation may involve iterative load steps in order not to overstress the cables) (Fig. 3.2). As a consequence of the application of cable forces, the deck may partially separate from the scaffolding. Alternatively, the scaffolding can be equipped with hydraulic jacks to lower the deck after the cable installation with the correct design length. Finally, the scaffolding is removed, and the cable forces are measured to allow minor corrections (Fig. 3.3). The application of superimposed dead loads (e.g., pavement and railings) may be executed either before or after removing the scaffolding, depending on the planning of the work. It would be preferable to execute these operations after

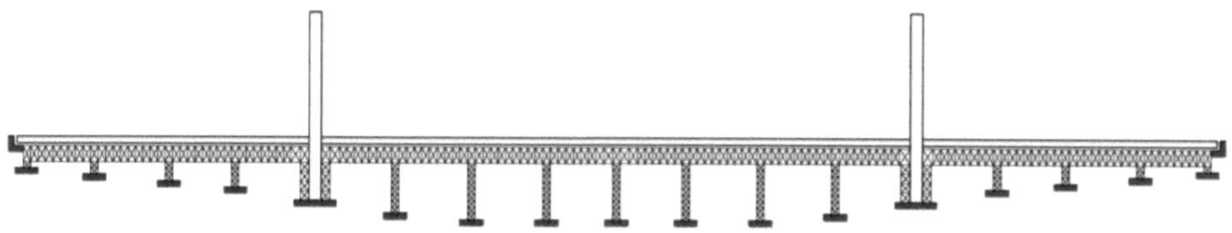

Fig. 3.1 Construction of abutments, piers, and deck

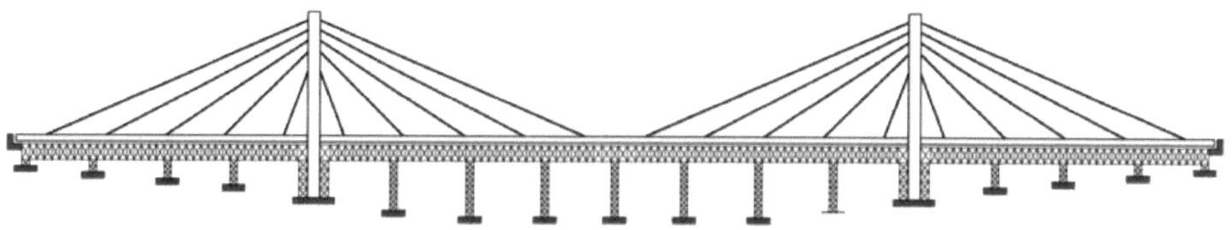

Fig. 3.2 Cables installation and stressing

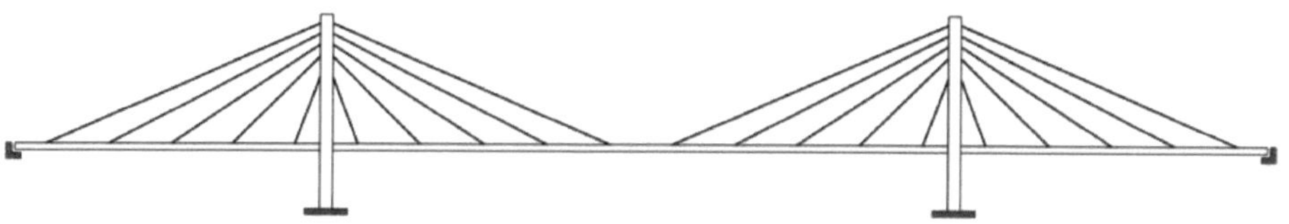

Fig. 3.3 Removal of scaffolding and cable restressing

removing the scaffolding to make the cable force adjustments more pronounced (and then easier to control) as will be explained further below.

As mentioned above, this type of construction method does not use the cable system as an element of bridge erection. In this case, the cable system is considered an ordinary element of the bridge which can be installed at practically any time. This rather begs the question as to whether the cable-stayed concept is really justified in a case where the deck can be erected by means of scaffolding. One possible justification could be the fact that the depth of the deck in a cable-stayed bridge is smaller than in a girder bridge and, consequently, this bridge type may be the only possible solution if the vertical clearance determines the design. However, in most cases, the choice of the cable-stayed solution is simply based on the desire to make an iconic structure in a particular urban or suburban landscape and no further justification may be found.

This principle of construction is also applicable to cases where the bridge is built at a temporary location and is then moved to its final location. This is something that has been done in cable-stayed bridges with a single tower by rotating the full bridge with respect to the vertical axis of the tower, such as for the Ben-Ahin Bridge in Belgium. It has also been applied quite frequently in other types of bridges like, for example, bowstring bridges.

3.1.2.2 Construction Analysis

The concept of construction analysis for this method of construction is very simple. Once the deck has been cast and, in the case of a prestressed concrete deck, it has been also prestressed, the deck profile is defined by the function $y_0(x)$ where x is the longitudinal direction of the bridge. This function defines the final position of the scaffolding and usually coincides with the designed vertical alignment of the bridge. In this respect, it must be taken into account that the scaffolding has a certain flexibility and its top surface may deform after casting concrete. This is particularly important in the case of scaffolding supported on discrete columns (an almost necessary condition in urban bridges to allow the passage of traffic under the bridge). Then $y_0(x)$ must be defined with a certain pre-camber and when the deck's cross-section is cast in different phases, prestressing must also be designed after taking into account the scaffolding's deformation under the weight of the deck.

Once cable forces are applied, the vertical alignment will be defined by a second function $y_1(x)$. As the deck may not be completely separated from the scaffolding, the exact shape of this function is very likely unknown. A possible solution to this problem would consist of slightly over-stressing the cables to force the separation of the deck from the scaffolding or under-stressing the cables to prevent this partial separation. The following step would consist of removing the scaffolding to achieve a new alignment, $y_2(x)$. As opposed to $y_1(x)$, $y_2(x)$ may be easily determined by calculations if the order of the construction phases is inverted: first, the scaffolding is removed and then the cable forces are removed. Consequently $y_2(x)$-$y_0(x)$ would correspond to applying the dead load of the deck to the completed bridge and then stressing the cables. In any case, it must be remembered that, as the separation of the deck and scaffolding is gradual, it is very difficult for the actual distribution of cable forces to be foreseen by analysis.

Finally, the superimposed dead load is applied on the deck to get a new alignment, $y_3(x)$, and cables are re-stressed to get $y_4(x)$, which is supposed to define the deck's final position at the end of construction. These two operations may be interchangeable. When creep and shrinkage are considered, $y_4(x)$ becomes $y_5(x)$, which corresponds to the defined design alignment of the bridge.

Although the whole process appears to be quite simple, in practical terms, the phases referred to above might get mixed up because other external constraints are also present: the need for urgent finalisation of the work to meet a given date for opening the bridge may drive the contractor to push for applying the superimposed dead load parallel to stressing the cables since this process takes some time. Another constraint consists of the fact that re-stressing the cables is also used to correct differences between measured values of $y_3(x)$ and corresponding computed values of $y_3(x)$. Another element which must be taken into account is obviously temperature. All these operations have been described for a uniform and constant temperature; any temperature variation would require appropriate corrections in the process. The different phases of analysis are summarised in Table 3.1.

3.1.2.3 Cable Force Application

Although bridge designers tend to talk about cable forces, as this is the parameter that defines the state of stress in the bridge, the process of imposing a cable force on any cable is quite intricate. In the case of construction on fixed supports,

Table 3.1 Description of construction phases for analysis

Phase	Vertical alignment	Vertical displacement	Loads for analysis
Initial phase after casting deck	$y_0(x)$	-	-
Stressing cables and removing scaffolding	$y_2(x)$	$y_2(x)- y_0(x)$	Deck dead load on deck only + stressing cables
Superimposed dead load	$y_3(x)$	$y_3(x)- y_2(x)$	Superimposed dead load
Re-stressing cables	$y_4(x)$	$y_4(x)- y_3(x)$	Stress variation in the cables
Final phase	$y_5(x)$	$y_5(x)- y_4(x)$	Creep and shrinkage

the analysis may be able to fully separate the two phases of cable stressing and scaffolding removal. In reality, though, these two operations are being performed simultaneously since the deck will gradually separate from scaffolding as cable forces are applied. Thus, it is almost impossible to know what force should be applied to a particular cable, as this load may depend on the cables that are already stressed and on the extent of the contact between deck and scaffolding.

One possible path of action could consist of applying a force smaller than the force needed to support the deck. In this way, it would be possible to assume that there is full contact between the deck and the scaffolding during the whole process of cable stressing. This is a delicate option as it must be verified that the deck can withstand its own weight without any damage once the scaffolding is removed. The alternative and opposite option would consist of slightly over-stressing the cables to be sure that the deck will separate from the scaffolding. A reasonable value of this excess cable force would be to reach the force needed to support the deck weight and the superimposed dead load. As deck separation from scaffolding would not be reached by stressing the first cable, the unknowns about the supporting conditions of the deck would still be present in this case.

Moreover, as cable force is very likely to be meaningless (i.e. difficult to evaluate) in all cases, the cable stressing operations should be done by monitoring cable elongations (and deck movements). This is the value that must be given to the cable supplier and taken into account in the first operation of cable stressing. After removing the scaffolding, cable forces must be measured to evaluate the possible errors and correct them in the cable re-stressing operation.

Re-stressing of the cables should be mandatory since it is meant to:

- Increase cable forces to support superimposed dead loads
- Correct possible errors during the first tensioning phase. It is assumed that the stress in all the cables can be measured at any time
- Correct irregularities in alignment which could not be corrected by playing with the position of the fascia beam. This is less important in the case of construction on fixed supports where it can be assumed that these irregularities may be very small.

Again, control of re-stressing through cable elongation is recommended as the cable re-stressing may coincide in time with other significant operations such as paving the bridge. It must be taken into account that, if a mono-strand jack is being used, only the cables which have to be elongated by more than about 50 mm can be re-stressed to avoid "biting " (i.e. physical penetration) in an already damaged surface of the strands. In this respect, one possible option would consist of stressing the shortest cables to their final stress during the first stressing phase and only re-stressing the longest cables.

Even if the re-stressing operation can be performed without any interference with other construction operations, it may be necessary to repeat re-stressing if it is found by measuring the stress in the cables that the differences between measured and computed values is relevant.

3.1.3 Construction by Incremental Launching

3.1.3.1 Construction Description

Incremental launching is a construction procedure that is well-developed for long girder bridges of a constant depth. It has many advantages including minimal interference with the landscape, the concentration of most construction work at a well-protected and secure location, the industrialisation of the construction work and the corresponding high precision of the resulting structure, and, finally, the higher standards of health and safety conditions which are achieved with this method. The drawbacks are mainly related to the limitations of this method since it requires specific horizontal and vertical alignment conditions as well as a significant area behind one of the abutments in which to concentrate most of the construction operations. This method has been applied to numerous girder bridges, especially railroad bridges, but it has also been applied to some bowstring arches.

In the case of cable-stayed bridges, there is a very well-known example that demonstrates that this method may be applicable to this kind of bridge: the Millau Viaduct in France with a total length of almost 2.5 km and span lengths of up to 342 metres. The cable-stayed scheme is particularly well-suited to be incrementally launched as the stays permit launching extensive lengths of deck even if the deck is very slender. Moreover, in the case of concrete decks, stay cables allow a large reduction of launching prestress. Nevertheless, two important drawbacks may be detected. The first one is related to the very high bending moments that may develop in the deck under the tower in certain phases of the launching process. This effect may be diminished by adding temporary piers and by reducing the weight of the tower and the cable forces. The second drawback is the substantial length of the prefabrication yard necessary (equivalent to one span length) since the stay system must be symmetric. This drawback may be just secondary for continuous bridges with many spans, as was the case for the Millau Viaduct. In more standard cable-stayed bridges, with one main span and two lateral spans, the incremental launching method may be more difficult to justify.

It must be mentioned that the cable-stay system may be used as an auxiliary means for launching a girder bridge to reduce the bending moments in the cantilevered deck. In this case, the launching process requires an adjustment of cable forces during the different phases depending on the position of the tip of the deck and on the position of the temporary tower in relation to the span.

Incremental launching for cable-stayed bridges permits different variants. One of

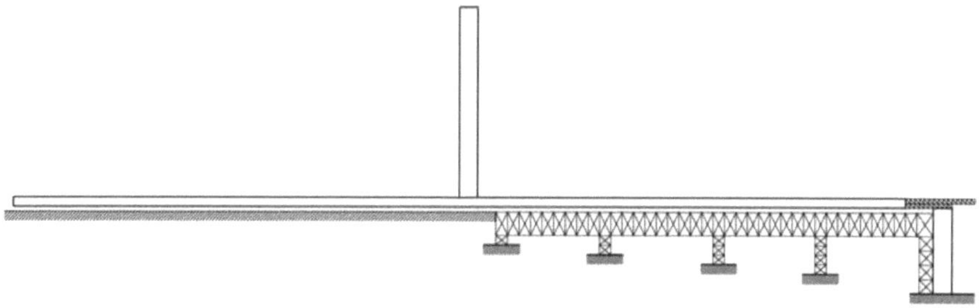

Fig. 3.4 Construction of one span length over a yard and scaffolding

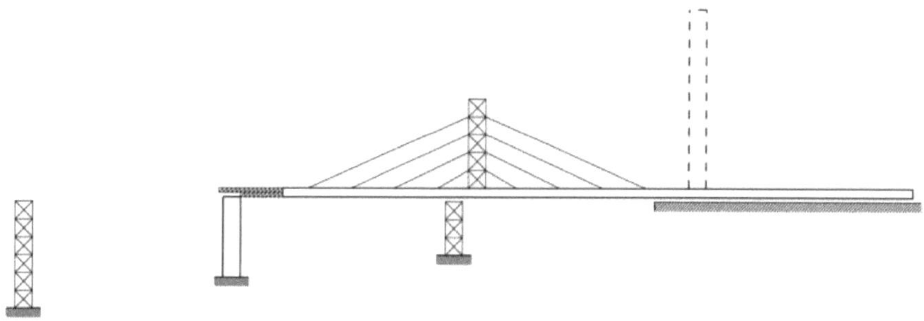

Fig. 3.5 Launching of deck with a temporary cable-stay system

the possibilities of this method consists on casting (or assembling and casting for a composite deck) one full span length in a yard adjacent to the abutment of the bridge or the lateral span (i.e. adjacent to main span) in the yard and half a main span over scaffolding as shown in Fig. 3.4. The tower must be incorporated with the deck and, as it must be launched with the deck, it may be useful to make it in steel instead of concrete, as was done for the Millau Viaduct. Concerning any incrementally launched girder, a launching nose may be incorporated at the front edge of the deck. An alternative to the previous system would consist of launching the deck with a smaller temporary pier to avoid the very long space (yard and scaffolding) reserved for building the deck.

In this solution (Fig. 3.5) a cable-stayed temporary tower is built with a smaller height than the final tower. The launching spans are divided by two by adding temporary piers at a distance of half the lateral span length. No scaffolding is necessary in this case. The tower may be installed on the deck already in the yard (if it is light enough) or it may be installed at its final location, above the corresponding pier to avoid its additional load on the deck. The advantage of using a temporary tower and a temporary cable-stay system is that the load on the temporary cables may be changed as needed by the deck without having to worry about the problems usually associated with alternately stressing and relaxing the cables.

For a classical 3-span bridge, Fig 3.6 shows two alternative systems. On the left, the full half of the bridge (with tower and cables, without a launching nose) is launched over its corresponding half span. The cable system may be able to maintain the deck within admissible bending limits

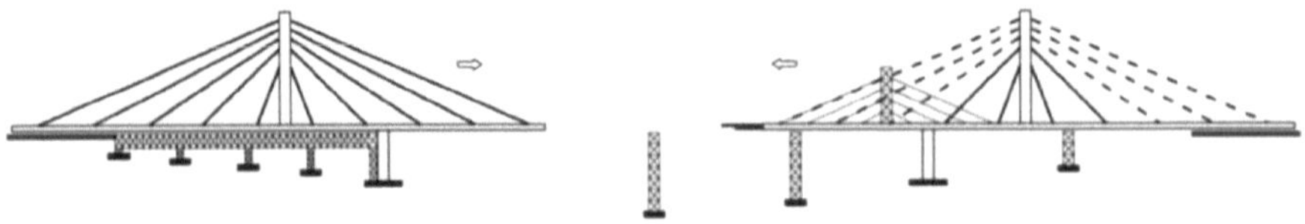

Fig. 3.6 Launching of deck over main span by two alternative systems

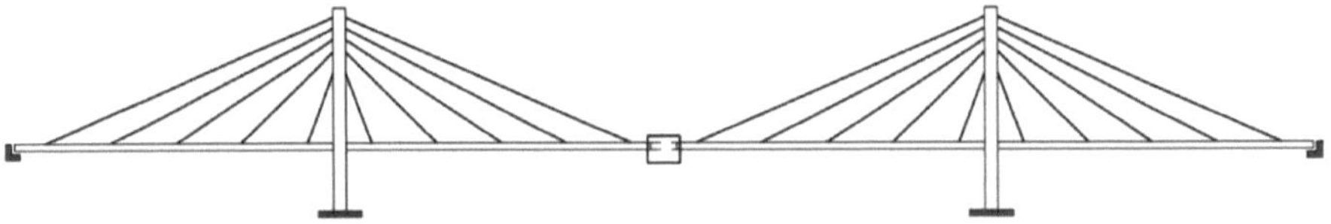

Fig. 3.7 Closure of main span

without the help of any temporary pier. As the lateral span (i.e. the span adjacent to main span) is supported by scaffolding, the full tower may be installed before the beginning of the launching process. On the right, the temporary stay system permits launching quarter-span lengths although it means changing the cable forces as necessary (the temporary cables would be stressed when the temporary tower is above a pier, either temporary or definitive). Behind the temporary tower, the final tower may be built (or not) and some of the cables stressed, although it would be preferable to wait and build the tower above its corresponding pier. These processes are very much influenced by the nature of the deck (concrete or composite) and the tower (concrete or steel). The distance between temporary piers is defined by the bending capacity of the deck without the help of any cable-stay.

The final stage (Fig. 3.7) consists of closing the main span after removing both launching noses, removing the temporary piers, and re-stressing the cables to accomplish the design vertical alignment.

The incremental launching method is generally not applicable to extradosed bridges for two reasons: the depth of the deck is generally variable for these bridges and the cable-stay forces do not fully compensate for the weight of the deck. This second reason could be obviated by means of prestressing, but this is usually expensive. Its use is illustrated in the cantilever method.

3.1.3.2 Construction Analysis

As explained above, many options are possible for the construction by incremental launching and will depend on many factors: type of deck and type of towers, length of the bridge, number of spans, span length, or even a possible disassembly of the deck during launching (launching the box without overhangs, launching the steel structure without the concrete slab). In any case, a great number of construction stages must be defined and the structure will grow during the process (elements are added to the model). Moreover, the supporting conditions will also change from one stage to the next (in the model supports are moving and the deck stays immobile while in the actual bridge, the opposite happens: the deck moves and the supports stay still).

As for launching ordinary girders, the deck deflects downwards while it is being launched and this negative vertical deflection must be corrected when the launching nose reaches a pier (temporary or definitive) by applying jacks between the deck and the pier. Hence, the deck deflections will oscillate during the launching process, which must be corrected at the end by re-stressing the cables.

The analysis of the construction process must consider all the intermediate stages with different structural configurations: different numbers of elements when new deck segments are being cast and different boundary conditions as the location of the supports will be changing relatively to the deck. Then, as opposed to the construction over fixed supports, a great number of different models must be prepared or, as most software packages do nowadays, elements must be added or removed during the follow-up of the construction process.

In the case of the supports, when a new one is added to the model or when it is moved to a new position, an imposed vertical displacement must be applied to it in order to force the condition that the deck has to maintain its alignment at every support. As this displacement is not known beforehand, the process becomes cumbersome since each step is only defined after knowing the results of the previous step.

The application of gravity forces must follow a two-step pattern. Before moving the deck, gravity forces must be applied with a negative value. Then, the model is modified by adding/removing elements and supports. Finally, gravity forces are reapplied with a positive value on the modified model. Depending on the software capabilities, these operations might be performed in a single step.

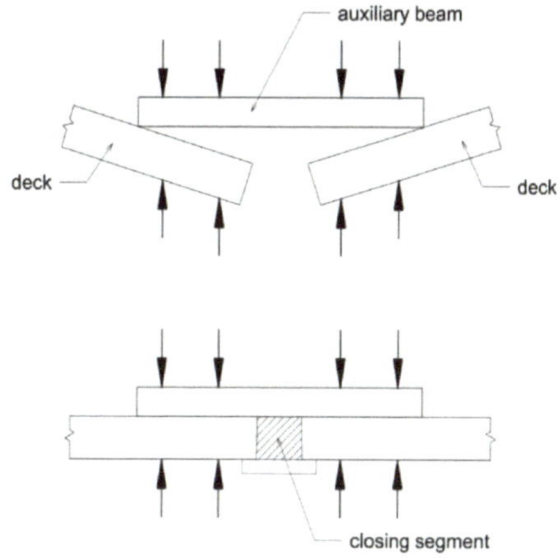

Fig. 3.8 Slope continuity at closing span

Closing the span is also a delicate operation. Even if it is assumed that the deflection is the same on both sides, which is not always true, slope continuity must still be achieved. This could be reached by a very precise pre-camber during construction, but it is almost impossible to reach full slope continuity before closing the span. Therefore, it may be necessary to force it by means of a mechanical device. This can be done by applying opposite bending moments on both sides of the cantilever using auxiliary beams and jacks as shown in Fig. 3.8. This device may also be useful to force displacement continuity as well on both cantilevers.

3.1.3.3 Cable Force Application

The definition of cable forces at every stage depends on the type of launching construction used. If the tower with all its stay cables is being launched as shown in Fig. 3.4 and the left part of Fig. 3.6, the cables should be stressed while the full deck is resting on fixed supports to a stress which exactly compensates the dead weight of the deck or slightly below this value. Accordingly, the cables should not be re-stressed until the end of the launching process. At this moment, if no intermediate supports have been used, cables should be re-stressed in order to approximately re-establish the vertical alignment of the bridge. After the superimposed dead load has been applied, the cables should be re-adjusted to compensate for the new loads as well as to reach the final design vertical alignment and to smooth all possible irregularities of this alignment (which are very unlikely with this construction procedure).

If intermediate supports are used during the launching process (as in Fig. 3.6), cable force adjustment before closing the span may not be necessary and it would only be necessary to re-stress the cables after applying the superimposed dead load.

If a temporary cable-stay system is used during launching (as in Fig. 3.5 or the right part of Fig. 3.6), the cables of this temporary system must be continuously adjusted as they must support the weight of the deck when the cantilever is large and they should be un-stressed when the tip of the cantilever is being supported by a pier (temporary or final). As this cable system is temporary, there should not be much problem in successive cycles of stressing and relaxing the cables, although some previous experimentation would be necessary to ascertain that the safety of this system is not compromised.

Again, the operations of final re-stressing of the cables and application of superimposed dead load may be interchangeable if the resulting bending moments in the deck are acceptable. As the contractor will push for executing both operations simultaneously, a very precise method of cable stressing by measuring the cable elongations and monitoring the deck displacements and cable forces will have to be set in place.

3.1.4 Cantilever Construction

3.1.4.1 Construction Description

Although initially proposed and applied on girder bridges, the cantilever construction method is ideally suited for cable-stayed bridges since the stay cables may totally or partially support the weight of the cantilever segments. This is valid, with obvious differences, for concrete, composite, or steel decks. The main advantage of cantilever construction consists of the possibility of building large spans without any temporary support in these spans. Therefore, it may be used to straddle across deep valleys, large rivers, or even sea straits without any interference with the ground, possible traffic, water currents, or ship traffic. Other advantages include minimising the footprint in the landscape, cost reduction due to the avoidance of large temporary works, minimising construction duration, and making construction relatively independent of soil conditions, weather, floods, or traffic under the bridge. Although construction on fixed supports is usually the preferred method for short-span and low-level cable-stayed bridges, cantilever construction is usually the preferred method for medium-sized and large-span cable-stayed bridges. This also holds for extradosed bridges.

In concrete decks, a form traveller is usually required to support the forms for the deck and the weight of fresh concrete until it hardens and prestressing is applied (and the forward cable is stressed). However, travellers are not necessary when concrete segments are precast or when the deck is composite or in steel. In these cases, segments are transported either along the previously built deck, to the ground located under the bridge, or by a ship when the bridge crosses a river pro-

Fig. 3.9 ▶ Description of cantilever construction

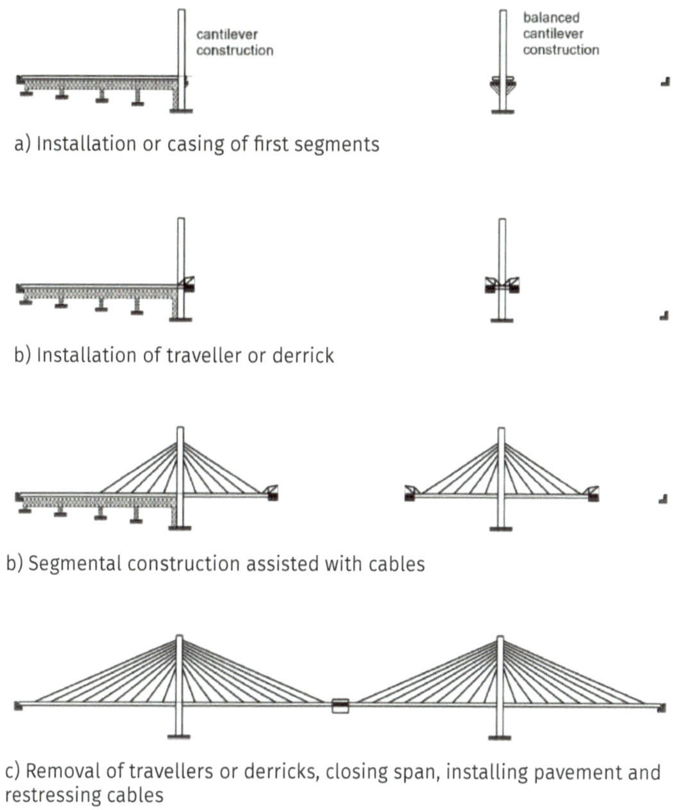

vided the depth of the river is sufficient. The segments are set in place by means of a derrick.

Depending on the distribution of spans and on the height of the deck relative to the ground, construction may be organised in two ways: simple cantilever construction or balanced cantilever construction. In simple cantilever construction, the side spans are built first by any of the possible means (on scaffolding, on fixed supports, or even by movable scaffolding or by incremental launching, although this is quite unlikely unless the length of the lateral spans makes this procedure economically competitive. Here, one or two segments of the main span must be built on fixed supports. Due to the height of the deck, these fixed supports are usually inclined struts which are supported on the tower (Fig. 3.9a). The length of these first segments is the minimum to allow the installation of the form traveller (Fig. 3.9b). For a single cantilever, the length of the first segment (usually named 'segment 0') is comparable to the longitudinal dimension of the tower. For a balanced cantilever, the length of the first segments is usually greater than the longitudinal dimension of the tower since these segments must accommodate two form travellers on top of them without any interference between them.

There is another important difference between simple cantilever and balanced cantilever construction. While in the first case, the solid connection of the first segment to the lateral span constrains the rotation of the newly built segment, in the second case, it is necessary to constrain the rotation of the deck. Then, either the deck is built solid with the pier or the provisional inclined supports must be maintained in place to prevent deck rotation.

After building the first segment and installing the form traveller or the derrick, construction follows a regular and repetitive pattern: casting or installing a new segment, connecting this segment to the previously built deck (by welding or bolting for steel or composite decks, by installing and stressing prestressing for concrete decks, either cast-in-situ or precast), moving the form traveller forward and installing and stressing the forward cable-stay (Fig. 3.9c). As the length of segments is limited to about 6 or 7 metres (for concrete decks) and the distance between cable anchorages in the deck may be larger, it may happen that the previously described sequence of operations is altered in the sense that the form traveller is moved twice between the installation and stressing of two consecutive stay cables.

For the balanced cantilever construction, when the length of the cantilevers reaches a certain limit if the verification of design unbalanced forces is not expressly imposed under the applicable design codes, the tower may not have the necessary strength to resist possible unbalanced forces from the deck. These unbalanced forces may be due to small differences in the moment of application of dead weight forces (it is practically impossible to build both cantilevers symmetrically), or due to the accidental fall of the traveller form on one side, or as a result of unbalanced lifting wind forces due to turbulence. Then it is necessary to support the deck on one side of the tower on a temporary pier (Fig. 3.9c) able to resist both tension and compression forces (which is usually called a push-pull prop). The connection of this pier to the deck must include elastomeric bearings and vertical prestressing to fulfil this double function.

In the case of steel and composite decks, the steel segments may be brought into position with a derrick from below (for instance from a barge if the bridge is crossing a navigable river) or moved along the lateral span up to the tip of the cantilever. If the deck is a box (an open box for a composite deck or a closed box for a steel deck), the segment is usually complete; but if the deck is a ladder-type structure (two edge beams and cross beams), the segments may be brought by pieces (detached partial elements) and assembled onsite. In the case of composite decks, the concrete slab may be cast in situ, although nowadays it is more usual to break it up into rectangular precast slabs which are supported on the longitudinal girders and cross beams. The longitudinal and transverse joints are cast in situ once the steel structure has been connected. The lateral edges of the concrete slab are generally cast in situ since the density of longitudinal reinforcement is higher in these areas and because there might be some interferences with other elements of the bridge (cable anchorages, parapets, drainage, fairings, lighting, etc.).

The final stage in construction consists of closing the span (Fig. 3.9d). The form travellers must be removed and only the forms for the closing segment are maintained. Although the biggest challenge during deck construction is to keep the design alignment, small corrections in vertical alignment may be applied by devices such as the one sketched in Fig. 3.8. Once the deck has been made continuous, the final two operations (application of superimposed dead loads, such as pavement, barriers, fairings, lighting, services, and re-stressing of stay cables) may be executed. There is no preferred order in the execution of these operations, but ideally, they should be performed sequentially. Nevertheless, at this stage of construction, there is usually some urgency in finishing construction and they tend to be executed simultaneously. As the actual response of the bridge does not coincide absolutely with what is predicted by computations, it is sometimes necessary to perform a final adjustment of cable forces.

Stay cables do not have to be stressed necessarily two or three times as previously described. As pointed out in the next section, they may be stressed just once if the whole process is reasonably coincident in terms of displacements and cable stresses with design predictions and if in this process the deck is always bearing the permissible stresses.

3.1.4.2 Construction Analysis

The construction process is especially delicate for cable-stayed bridges which are built through the cantilever method since the deck supports very high opposite forces: dead weight forces (of the deck and of the form traveller) and cable forces. This process must be designed with the aim of avoiding damage in the deck (through excessive stresses or cracking) and cables. The standard calculation is based on inverting the construction operations and it includes the following steps:

- The final distribution of cable forces must be defined by the designer. It is usually such that the bending moment distribution in the deck is the same as for a continuous girder which would be supported in vertical direction at the anchorage points of the stay cables. This method cannot be applied in the same way if the bridge is not symmetric with respect to the tower; in this case, it may be more important to minimise the bending moments in the tower. There are alternative ways of defining the final cable forces, but they may not be considered a part of construction analysis. This final situation is supposed to correspond to infinity (t_∞).
- The creep and shrinkage loads are applied with a changed sign to define the situation at the end of construction, t_{ec}. As creep and shrinkage cannot be analysed backward, the forward analysis of construction must be used to invert the process
- The next step consists of removing the superimposed dead loads (pavement, barriers, etc.) and relieving the cable forces if it has been foreseen that re-stressing would be necessary at this stage. A feasible and simple criterion to define these force decrements could be to maintain the deck shape reasonably close to the design alignment
- Then the closing segment must be removed, and the corresponding forms must be removed (before removing the segment) and added (after removing the segment). After this step, the bridge would be divided into two independent parts
- The next step consists of applying the form traveller (or derrick) forces. At this stage, the analysis has reached the situation at the end of the repetitive process described

above (see "segmental construction assisted with cables" in the Fig. 3.9). Then, the same process must be repeated in the reverse order:

- The front cable is unstressed and removed. The cable force before relaxing the cable is an important result since this is the force that must be applied to this cable during construction at this stage
- The front deck segment is removed
- The form traveller is moved backwards one step
- If there are two segments for each cable, the next deck segment is removed, and the form traveller is moved backwards a further step.

- At the end of this repetitive process, the last segments (or the first if the order of construction defines what is understood as the first segments) are removed
- Finally, the form traveller is removed and segment 0 is also removed.

This simple scheme can be completed if prestressing operations are included in the case of concrete decks or if precast slabs and casting of the joints are included in the case of composite decks. An important datum in these calculations is the weight of the form traveller (or the derrick). In the preliminary (and final) design of the bridge, this weight is unknown, and an approximate value is adopted: a value around 65 % of the weight of one segment is generally a good approximation. During construction, the actual weight of the traveller is known, and the calculation of the whole process must be repeated. It must be understood that the weight estimation is approximate since the weight of the traveller depends on the type of cross-section and on the fact that it corresponds to the full section or only a central part of it (for instance, the central box of a section with overhangs).

An effect which may be very relevant in many cases is the influence of concrete deformation at early ages, conditioned by creep, shrinkage, and the hardening process of concrete. Another relevant effect, especially at early stages, is the fact that there may be differences between the actual weight of concrete segments and the theoretical weight according to design. This weight may vary with respect to computed values due to slight variations of dimensions or because of variations of the concrete density, or arising from an incorrect prediction of reinforcement density. When these weight and stiffness variations are compared with theoretically obtained cable forces, the resulting deck deflections may be different from computed values. Therefore, the reason for the differences must be investigated and the cable forces must be modified accordingly.

In any case, the methodology for construction analysis depends heavily on two circumstances:

- The relative stiffness ratio between the deck and the system of cables. A very flexible system of cables would mostly rely on the deck and the construction process would be very similar to that of a girder without any stay cables. The bending moments would grow parabolically from the tip of the cantilever towards the main pier. In a cable-stayed scheme, the combined action of the cables greatly reduces the cantilever bending moments and makes the construction system more efficient. At the limit, for a very stiff system of cables, the only cantilever would be the part of the deck located in front of the last cable. In practice, the actual situation is somewhat in between these two extreme cases.

- The amount of re-stressing applied to the cables in the last phases of construction. If no re-stressing is foreseen, and apart from a greater uncertainty about the possibility of correctly adjusting the final alignment of the bridge, the initial stressing force may be too high. On the other hand, if re-stressing is foreseen at the final phases of construction (usually in combination with the application of the superimposed dead load), the stress of the cables and the bending moments in the deck may be controlled to prevent excessive forces and moments.

Although the analysis of bridge "deconstruction" is necessary to define the different stages, only forward bridge construction analysis gives all the answers: the state of stresses in the deck, deformation and deflections of the deck, cable forces, and delayed deformations. The arrival at the design vertical alignment is a very essential issue which can only be fulfilled by a continuous control of deck deflections and a readjustment of camber (both for concrete and composite decks).

3.1.4.3 Cable Force Application

For most cable-stayed bridges, an important constraint consists of minimising the bending moments in the towers. Thus, cable forces should generally have the same horizontal components for the main span and the back span. Obviously, this does not apply to inclined towers nor to non-standard designs where the number of back span cables is reduced to a minimum or even to zero. In many other cases, the zero-bending moment in the tower condition is not fulfilled on a cable-by-cable basis but on an overall basis; this happens, for instance, when back cables are concentrated around back-span supports.

The front cable of the cantilever can only be stressed when the form traveller leaves the necessary space to do it; this depends on the position of the cable anchorages (at the centre or at deck edges) and on the type of equipment (form traveller, under deck or above deck, or derrick) and on the type of deck (concrete, steel, or composite). In any case, the unsupported part of the cantilever is subjected to hogging bending moments which may be compensated by cable forces. As shown in Fig. 3.10, the typified bending moment diagram reaches a maximum after a few segments to diminish afterwards, which is a completely different behaviour from that of a girder bridge without any stays. The amplitude of this bending moment diagra (both in moments and in horizontal length) depends on the relative stiffness of cables and girder as well as on the type of material, as mentioned above. This limits the amount of construction prestressing to be installed in the top slab and it makes it relatively constant along the deck. Consequently, it is frequently convenient to use prestressing bars, which may be connected easily.

Cable force application shows an interesting behaviour. When the front cable is stressed, the cables located behind are slightly unstressed, so the cable force diminishes along the whole cantilever construction. The cable forces will only increase at the end of the construction process when the superimposed dead load

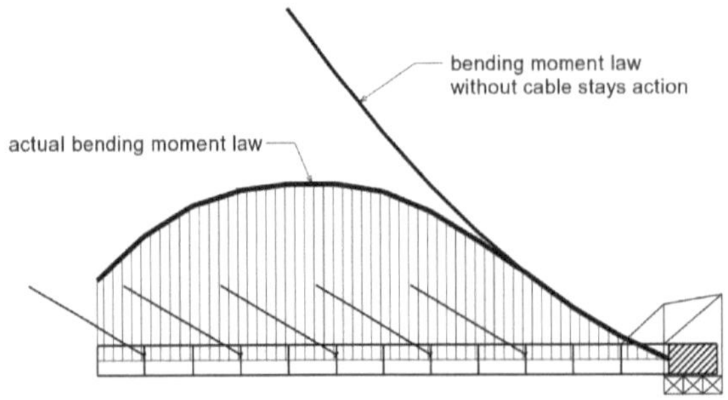

Fig. 3.10 Typified bending moment diagram during cantilever construction

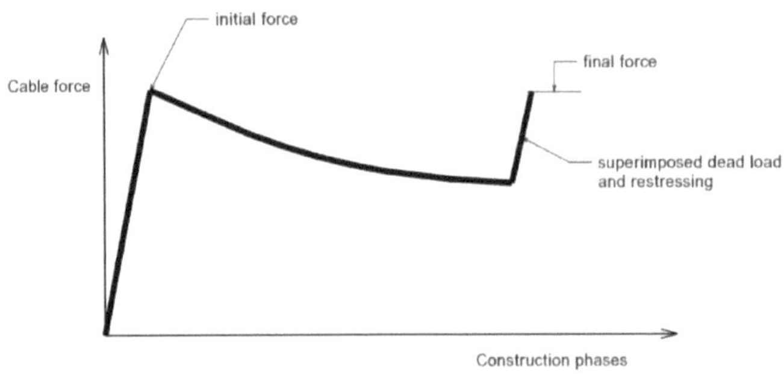

Fig. 3.11 Cable force variation during cantilever construction

is applied and the cables are re-stressed if necessary. This variation is shown schematically in Fig. 3.11 (actual variation is not as smooth as shown due to the different consequences of the operations involved such as form travellers' movements, casting segments and longitudinal prestressing, or, in the case of composite decks, installation of precast slabs and casting of joints). The shape of this diagram has some important implications since, ideally, the final force should be close to the limit load (around 0.4 floating point units or fpu); then the initial stressing force might be greater than the service limit load. This is allowed in most structural codes, but the cable stress must be checked at all stages as the fulfilment of this condition may have some consequences in the construction process.

This fact also points to the importance of re-stressing at the end of construction. If the cables are stressed at the level just necessary to compensate for the weight of the deck (to avoid overstressing the deck), it is likely that the end stress of the cables will be excessively low to take advantage of the full capacity of the cables. Moreover, as has already been mentioned, cable re-stressing is also a way to make small adjustments in the vertical alignment of the deck.

3.1.5 Cable Installation

The cables represent the core technology in cable-stayed bridges. Most of the advances in these bridges have come from new developments in cable technology (and most problems and failures have their origin in the cables). Today it is possible to affirm that the technology of cables has reached a point of maturity and that the cables are very reliable. If only present-day technology is considered, there are two types of cables: locked coil cables and parallel strand cables (or prefabricated parallel wire cables).

Locked coil cables usually come from the supplier fully complete, with the designed length and their corresponding anchorages, either with hammerheads or with fork sockets. Therefore, the design must define the length of the cables with precision. This is quite easy since it is only necessary to consider the final geometry of the bridge and the final force on the cable; from these two data, it is possible to derive the length of the cable at fabrication (with zero stress). Nevertheless, this would be valid in a perfect world where the final geometry of the bridge would be the same as that designed.

There are two solutions to take into account small errors in construction. One of them consists of using adjustable anchorages, which are offered by most suppliers. The other one would consist of preparing the structure to allow for small variations in the position of the cable anchorage. Fig. 3.12 shows a solution designed for the Manzanares footbridge in Madrid (Spain) using long slotted holes for connections with high-strength friction grip (HSFG) bolts.

The installation of the cables is usually performed in small bridges or footbridges by lifting the tower anchorage to its anchorage point by means of a crane. For longer and heavier cables, it may be necessary to use a highline cable and to pull the tower end of the cable from the tower; the cable is connected to the highline cable while it is being deployed from the reel.

However, the most widely used type of cable is based on the use of parallel strands. These cables are easier to transport since all their components are transported onsite independently and the cable is mounted onsite; the system of installation is almost independent of the size of the cables.

The first step is the fabrication of the polyethylene external sheath. The sheath comes in segments which are deployed on the deck and welded on site. A lifting collar is attached to the upper end of the sheath, and this end is lifted to the tower anchorage by means of a pulling rope (or the first strand – master strand). Then the individual strands are introduced into the sheath and pulled up to the tower anchorage. When the first strand is stressed, generally to a low stress, the sheath reduces its sag to a value which is already close to its final sag.

The installation of the strands is relatively simple since only small units should be used at each step: individual strands and monojacks. The goal of the installation is to stress all the strands to the same stress. But it is obvious that every time one strand is stressed, the tower and the deck get slightly deformed, and the previously stressed strands reduce their stress. Therefore, the first strands should be slightly overstressed. Accordingly, the suppliers use a system called 'isotension', patented by Freyssinet, or their own system which is based on similar principles. The isotension method basically consists of installing a load cell in one of the first strands and stressing all the strands to the load indicated in the load cell. Alternatively, it is possible to pre-calculate the stress to be applied in each strand and at the end to make adjustments. Anyway, at the end of the process, all the strands should bear the same load. As already noted, this stressing process is usually performed in several steps.

An important fact to be taken into account is that at every stressing step, the jaws should not "bite" (penetrate) the steel on a surface which could have been damaged by

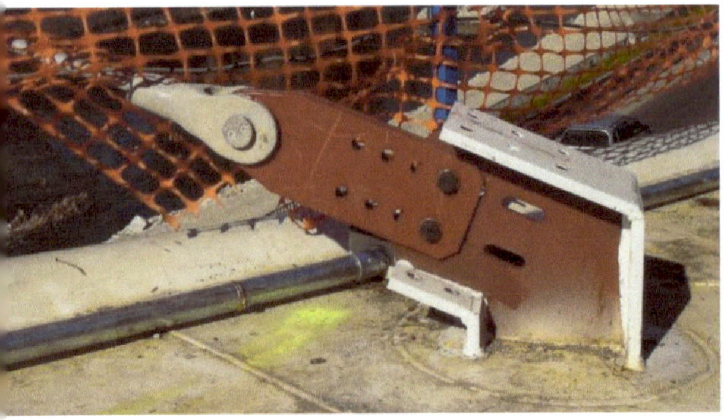

Fig. 3.12 Design for adapting cable length to actual bridge geometry

the jaws in a previous stressing step. This means that the strand elongation at each step should be longer than 40 to 50 mm. This condition is relatively easy to fulfil during cable installation, but it may be more difficult to accomplish during the final restressing. This means that restressing may in some cases be performed with a multi-strand jack, which is very heavy, and more difficult to move and place in position.

The description which has been presented for the two types of strands reveals that there are two types of installation data: by geometry or by force. These two types may generally be associated with locked coil cables and parallel strand cables respectively, but this association is not general: locked coil cables may be stressed by force, and parallel strand cables may be stressed by geometry. Stressing by geometry requires very strict geometric control; nevertheless, a 5 mm error in the length of the cable for a short cable (50 metres) would represent a stress error of 20 megapascals (MPa), so less than 3 % error in the load, which may be an acceptable figure. In any case, generally, designers prefer stressing by force because this parameter better relates to design computations and force distribution in the bridge.

3.1.6 Lessons Learnt and Recommendations

- In cable-stayed and extradosed bridges, geometric control and the retensioning plan are absolutely key issues. It is recommended that all intervenient in this process (designer/constructor/suppliers) should engage as a 'single team' with a very clear and thorough single plan comprising very clear specifications on 1) physical definition/marking of points object of measuring; 2) instants of measuring; 3) additional measuring context variables; and 4) with commonly accepted criteria for the interpretation of results
- In single-line cable-stayed and extradosed bridges where bolted steel shores and struts are used in the box girder, relative displacements on connections are to be evaluated and accounted for in geometric control
- Comprehensive three-dimensional building information modelling (BIM) is highly recommended in the first deck segment over the piers, comprising all structural elements and if applied, any inserts of form traveller or other erecting equipment
- In cable-stayed and extradosed bridges erected by the cantilever method, if it is structurally acceptable to move the form traveller prior to the 'last cable' tensioning, that option will significantly increase productivity (reducing production cycle lasting)
- In cable-stayed and extradosed bridges erected with falsework (fixed supports) sliding supports may be required.

References

[1] R.Walther, B. Hourlet, W. Isler, P. Moïa, J.F. Klein, Cable-stayed Bridges, Thomas Telford, 2003
[2] H. Svensson, Cable-stayed Bridges, Ernst & Sohn, 2012
[3] N.J. Gimsing, C.T. Georgakis, Cable Supported Bridges, Wiley, 2021
[4] PTI. Recommendations for Stay Cable Design. 6th Edition. 2012
[5] SETRA Design Guide. Prestressed concrete bridges built using the cantilever method.2007
[6] IABSE, SED17 on 'Extradosed Bridges'.
[7] AASHTO. Guide Specifications for Wind Loads on Bridges During Construction. 1st Edition 2017
[8] ASCE Guidelines for the Design of Cable-stayed Bridges, 1992
[9] J. Strasky, The Power of Prestressing, Structural Concrete, vol. 4, no1, 25–43, 2003

3.2 Suspension Bridges

Torben Forsberg[1] | Yasutsugu Yamasaki[2]

[1] Denmark, [2] Japan

3.2.1 Introduction

While most of the construction works of a suspension bridge are identical to those of other bridge types, there are a few major differences. First of all, the erection work of the main cables, cable clamps, hangers, and bridge deck sections is unique to a suspension bridge. Secondly, its abnormal proportions with very long spans and high towers create demands not experienced with other bridge types. Each of the various components must be strong and stable to withstand environmental effects such as strong winds, rough seas, unpredictable dynamics, etc. How can construction accuracy be assured with such a large structure? A bridge with a small span slightly outside geometric tolerances can be adapted onsite, but in the middle of a large span suspended high up in the air, there is no such room for minor adjustments. Thus, suspension bridge elements must be geometrically perfect. Logistics become crucial as there is only one way to construct the bridge with one sequential order of construction and no option of accelerating works, nor to do them in parallel. There is virtually no reliable Plan B.

The conditions on site are quite often harsh. Therefore, many elements of a suspension bridge have been prefabricated at the factory or workshop prior to arriving at the bridge site. Closed steel box bridge decks are prefabricated in full sections, truss type decks may likewise be preassembled into full sections as elements are transported and erected from water. Strands for main cables may also come prefabricated from the factory. Hangers, clamps, saddles, and main cable anchorage items are likewise made in steel shops and arrive fully finished at the site. If the towers are constructed of steel they are subdivided into rather large, prefabricated elements. Regardless of the amount of prefabrication, the building of a suspension bridge requires a substantial portion of the work to be carried out on site. This not only includes work directly related to the installation of the numerous elements but very significantly also all the preparatory works required to construct the many temporary working platforms and access walkways required.

The slenderness makes a suspension bridge deck susceptible to various vibration phenomena, and weather conditions become highly important with strongly pronounced wind effects. Many suspension bridges are built across water, and some could be considered offshore structures, therefore, adverse sea conditions are another challenge to meet. With size also come quantities and repetition of each work task. It is important that there is no systematic mistake or deviation in any work sequence, otherwise, a severe drawback in terms of production rates could result, and in turn a major economic loss.

It takes a long time to build a suspension bridge, so unpredictable seasonal weather variations make the construction phase of a suspension bridge very difficult. The main cable construction can be planned to be carried out under calm weather conditions during summer, but due to the knock-on effects of a delay of one of the other parts of the bridge, the cable works may be subjected to strong storms or very cold weather during winter, unless the works are paused for a time. With a typical construction time for a major offshore suspension bridge of approximately 5 years, it is often the case that it may take a year longer to build.

This section mainly focuses on describing and explaining the essentials of construction methods applicable to suspension bridges. For a more in-depth analysis of this type of bridge, an extensive bibliography is available, as shown in the list of references at the end of this section [1–6].

3.2.2 Large Scale Effects

Suspension bridges are applicable where long, free spans are required, with a recent tendency for increasingly longer spans. The large scale of the elements leads to the generation of stronger forces supported by the foundations (anchorage and tower), in turn requiring sound geological conditions, sometimes large excavations, and leaving large footprints. The consequence is towers 200m high or more, and a heavier and longer main cable, which requires heavier transportation, large capacity cranes, and equipment for erection.

3.2.3 Weather Challenges

Wind is the most common challenge for suspension bridge construction. A severe storm is obviously an event that the temporary structure must withstand, but downtime due to various frequent wind phenomena may be even more troublesome. For example, lifting operations may often be temporarily suspended due to unacceptable wind gusts. Typically, this leads to a slower construction speed which must be incorporated in the project schedule. In a storm, the wind turbulence is relatively small, so to some extent vibration phenomena are more pronounced at moderate wind speeds, i.e. they are more frequent. No wind would mean perfect construction conditions for a suspension bridge. Unfortunately, it is virtually always windy on a large bridge.

Rough sea conditions are another challenge for suspension bridges, mainly for the project time schedule. Delivery of material and even workforce to the tower or anchorage foundation may not be possible, and marine work using barges and floating cranes for erection may likewise be so difficult that bridge design alternatives are chosen.

In colder climates where the temperature can drop near to or below the freezing point work can be very slow, risky, or impossible. On the other hand, high temperatures or large temperature variations make setting out and geometry measurement less accurate, i.e., it reduces the perfect accuracy of the structure that is otherwise required.

3.2.4 Offshore Works

Since suspension bridges are often built over the sea or a wide river, the construction of foundations (if these are built in water), towers (if made of steel), cables, and suspended decks installed from barge and floating crane are typically classed as offshore works.

For a suspension bridge over water establishing worksites can take a lot of time.

It could take a year until the first temporary platform is in place. Other activities can also take time, such as workers being transported to and from work sites, as it takes a long time to sail from one tower to the other, and so on. The worksite is often split in two, the main worksite on one side of the navigation channel, and a secondary site on the other. The split will generally reduce transportation time for material and workers. The work procedures must be well organised, as a split worksite implies additional demands in terms of logistics. Otherwise, this will have a negative impact on construction speed and project schedule.

3.2.5 Tower Construction

There are two types of towers, reinforced concrete towers and steel towers. Tower foundations are located offshore in many cases for long-span suspension bridges. Therefore, the use of a steel tower could shorten the construction schedule as prefabricated tower segments could be ready at the workshop for installation when the tower foundation is completed. This, however, would not shorten the schedule as expected if weather conditions are not favourable for the prefabricated tower segment lifting operation. Within steel towers, there are two types of connections of skin plates between segments, welded connections or (tension or friction-resistant) bolted connections. For the latter case, segment ends are machined for geometric control purposes. Steel towers with bolted connections could shorten the construction schedule further as the geometry is confirmed at the workshop by machined ends. Joining work could also be made quicker by only tightening bolts on site, however, this would cost more and require more time for fabrication. In cases where the offshore construction of the tower foundations does not take a long time or if the construction of the tower foundation is performed onshore, the previous statements do not apply, and a reinforced concrete tower is just as competitive.

Should a steel tower be chosen, accurate fabrication at the workshop is paramount for controlling geometry and misalignments on site. The accuracy of fabrication is confirmed at a trial assembly at the workshop before delivering each segment. It is preferable that the trial assembly is constructed in the vertical direction identically to how it will stand on site during erection although this can be difficult due to the weight and/or sectional dimensions of segments. In this case, the trial assembly should be constructed in a horizontal direction, with special care taken to prevent the effect of deformation.

The construction of a concrete tower is relatively straightforward. With a jump form, a construction speed of 4–6 m per week can be achievable, but connections to cross beams and completion at the tower top take longer as these works are more difficult. Alternatively, tower legs may be slip-formed, which has a potential time and cost-saving effect. Fig. 3.13 shows the Hålogaland Bridge under construction. The tower legs above the deck were slip-formed. Offshore tower construction typically requires a concrete batching plant at the tower base if transport time would otherwise be too long for delivery of fresh concrete.

3.2.6 Main Cable Erection

Main cable erection consists of two major activities, catwalk erection and strand erection. Parallel wire strand is normally used for long-span suspension bridges and

Fig. 3.13 Hålogaland Bridge, Norway, 2016, (TorbjørnS / Wikimedia Commons, CC BY-SA 4.0)

is constructed either by means of aerial spinning or it can be prefabricated. The catwalk design is generally the same either for a strand constructed by aerial spinning or a prefabricated strand, but catwalks used for the prefabricated strand method become heavier as they must support the strands as they are pulled from one anchorage to the other.

Catwalks can be designed with catwalk strands either continuous or terminated at the tower top. For a longer bridge span, it is preferred that the catwalk be designed with the catwalk strands terminated at the tower top. This restraint is useful for the strand sag adjustment which is achieved by sequential measurement and adjustment of the length of the strand in each of the three spans. The catwalk strands are anchored to the anchor girder which is connected to the tower top by means of long tie rods, enabling length adjustment and release. The tie rods and their connections should be designed very carefully, allowing for vertical rotation under live loading of strand hauling and transverse rotation under wind load, so as not to create excessive bending stress in the tie rod.

Prefabricated parallel wire strands are preferable as the span becomes longer since it is less sensitive under wind and could shorten erection duration. The strand comprises 127 wires being hauled by each operation, however at a speed lower than that of aerial spinning with 2–8 wires. Parallel wire strands are fabricated in a factory where wire length and absolute length should be accurately controlled so that the variation of tension among wires within the strand is below the acceptable level and the installation of strands at the site can be performed without any difficulty.

The construction of the main cables and the subsequent deck erection requires a full step-by-step analysis that always follows the geometry, the geometric change caused by load increase, the effect of catwalks and pull-back cable restraints, and temperature effects.

3.2.7 Saddles and Anchorages

The tower saddle of a suspension bridge typically has an upper cast steel trough on top of a box-type plated steel structure. The saddle may be relatively small, but it has a substantial weight and cannot be erected by normal tower cranes.

The trough's inner sides and bottom face are located against the main cable and must therefore be geometrically perfect. Each individual wire of the main cable must remain in the same perfect alignment across the entire length of the saddle. Regular or standard steel construction tolerances would not qualify; hence these faces of the cast steel must be machined. Spacer plates between the wire bundles or strands fulfil the same perfect geometric tolerances. It is also possible to construct the upper part of the saddle by welding thick steel plates and ensuring their geometric fit with machined fill plates.

To allow more options for bringing the saddle to the tower top, the structure may be subdivided into two pieces that are bolted together on site. To make the strand erection stable without slippage over the saddle, the tower saddles need to be slightly offset during the main cable construction. This can be achieved by either tower pull-back (tower top pulled towards the anchorages by means of temporary cables), or by saddle shifting. By saddle shifting, the saddle travels a short distance on the tower top and finds its final position centrally.

The splay saddles also have an upper trough which must be machined in all three dimensions to perfection, so that all main cable strands are diverted geometrically towards the anchorages in the bottom of the splay chamber.

The main cable strands are anchored to the crosshead slab which is anchored into rock or cast into the anchor block concrete, in many cases by means of post-tensioning tendons. Normally two or four strands are anchored to one crosshead slab for the main cable constructed by prefabricated strands. Therefore, the direction control of the crosshead slab is extremely important for not creating additional bending stress in the strand anchoring tie rods.

3.2.8 Installation of Cable Clamps and Hangers

Cable clamps are typically cast steel constructions. Cable clamps are horizontally split into two halves, where hangers are pin connected, or vertically, where hangers are mounted over the clamp depending on the type of hanger rope. They are attached to the main cable by high-strength bolts stressed to high loads, that effectively pull the two halves of the clamp together to fit with the cable geometry.

Prior to the cable clamp erection, the bundle of erected strands is compacted into a circular shape. This operation should be carefully controlled at the locations where clamps are to be mounted, in order not to over-compact the cable, in which case the clear gap between the two halves would close. The air void ratio (cross-section area of voids between wires/gross section area of cable) is approximately 20 % based on past experience with air spinning. The void ratio is smaller for a main cable with prefabricated strands, and the cable clamp is thus designed for those reduced void ratios (16–18 %). Prefabricated strands come perfectly compacted in a hexagon shape, and over-compaction might end up with an air void ratio lower than planned, requiring additional thin shim plates or dummy wires between the clamp and the main cable, to keep the gap for tightening of bolts also in the future.

Hangers are prefabricated to a very accurate length. The most common connection is by inserting a pin between the hanger socket and the cable clamp at the top, and the same at the connection to the bridge deck at the bottom. There is no adjustment possibility as was required for old suspension bridges. In some cases, the hanger is bearing type anchored (where the hangers are connected through bearings), where the length can be fine-tuned with the adjustment of shim plates.

3.2.9 Bridge Deck Construction

The bridge decks of modern suspension bridges are typically closed steel box girders. The steel boxes are fabricated in the

Fig. 3.14 Osman Gazi Bridge, Turkey, 2016, (Enisfesci / Wikimedia Commons, CC BY-SA 4.0)

workshop and as suspension bridges are usually located close to water, the segments are shipped to the site just in time to be lifted in place either by a floating crane or by lifting gantries on the main cable.

The exception to closed steel boxes is the truss-type construction used when the bridge deck is for two-level traffic, or if step-by-step erection starting from the towers may be preferred due to the reduced impact on ship traffic underneath.

Regardless of the type of bridge deck, the top of the deck is orthotropic formed by a deck plate and U-trough that is very much subject to fatigue, therefore it should be carefully designed and fabricated for a pre-determined permitted discontinuity. Geometric control during fabrication is important so that erection on site can take place without difficulty.

The deck segments of a steel box are weld-connected for perimeter plates and weld- or bolt-connected for a U-trough.

A floating crane is usually used to erect deck segments where a lifting gantry is not an option, e.g., near towers or at the far end of side spans. The deck segments are lifted, connected to hanger ropes, and temporarily connected to the previously installed segment, until the deck geometry allows for permanent connection by welding and bolting. The design of the temporary connections becomes more complex as the span gets longer because deformation is larger under wind and vertical load. The dead weight of the lifting gantry significantly impacts the design of the temporary connection if it is heavier with the strand jack system. The deck segments are intentionally not connected along some joints so as not to experience a large buffeting response in turbulent flow but feature hinged connections and larger gaps to prevent them from clashing into each other.

Fig. 3.14 shows the Osman Gazi Bridge during deck erection, which was predominantly carried out by means of gantries lifting the elements from the main cables.

3.2.10 Vibration

Divergence vibration is always the critical issue for the bridge deck for long-span bridges, but rarely for the towers. A buffeting response in turbulent flow occurs on any structural element. Wake buffeting occurs on hangers even in smooth flow by wakes created by other elements that are present upwind. Vortex-induced vibration occurs in smooth and even in turbulent flow on the bridge deck, the tower, the hanger, and the hand rope.

This vibration should be carefully investigated using a wind tunnel test and mitigated where necessary to below the permissible amplitude for safety against yield and/or fatigue, for human comfortability during in-service conditions, and for workability during construction.

The tower, if it is a slender steel construction, will in most cases experience vortex-induced vibration during construction, and even in the service condition, regardless of the cross-section of the tower being shaped to minimise vortices. This vibration is harmful to construction equipment safety and workability during construction and to structural safety, fatigue damage, and operationality in the service condition. Thus, this vibration should be mitigated which is achieved by using a (passive or active) mass damper. Where the natural frequency of the structure is very low and the target amplitude (acceleration) is very small for the tower during construction, an active mass damper might be the only option.

The cable structures, both the permanent cables and the temporary cables during construction, will tend to vibrate to a certain degree, most often initiated by wind directly or indirectly. Vibration can lead to fatigue deterioration, either of the cables or of items supporting or clamping them. Vibration can be at low wind speeds, a vortex type of vibration, but can also be at higher wind speeds where the effect can be more harmful, thus it should be mitigated wherever necessary. Upon completion, the hangers can vibrate, which may be technically of little concern, but this can create discomfort, hence mitigation means should be applied shortly after completion of the construction.

The bridge deck should be designed carefully by the adoption of wind tunnel tests in order not to experience any vibration of divergence type or the vortex-induced vibration that occurs more often. Once it happens after the completion of the bridge, it is extremely difficult to mitigate the situation unless the deck section is modified to change wind flow around it. Mass dampers cannot be expected to mitigate vortex-induced vibration, as frequency is very low, vibration occurs in a vertical direction, and space in the deck section is limited, restricting the possibilities of installing effective mass dampers.

The various vibration phenomena arise due to different sources or combinations of conditions which are difficult to forecast. Thus, when building a suspension bridge the various effects must be addressed beforehand, and mitigation planned so that these can be quickly implemented if necessary.

As many suspension bridges are old, minor or major degrees of retrofitting have proved necessary, to avoid complete demolition of the bridge. There are examples showing that major load-carrying elements like the main cables and bridge decks can be replaced. At times, even large replacement work is carried out while still allowing traffic on the bridge. Retrofit works on suspension bridges are to be a key aspect since the early stages of preliminary design and or tender conditions.[1]

---

1 Retrofit works on suspension bridges are to consider the following aspects: Hanger replacement is a quite common retrofit activity. Typically, various temporary structures are required, for example, a temporary hanger may be required while the old hanger is being substituted. Modern suspension bridges may also have been designed to allow for hanger replacement to take place without a replacement hanger, i.e., the increased temporary loads to neighbour hangers are acceptable.
Deck replacements or major deck repair works necessitated by fatigue wear are also possible retrofit works. Many bridges have been furnished with dehumidification systems for the main cables and closed steel compartments after they were built and not originally designed for this.
As retrofit works are further constrained when compared with initial construction, there are examples that such works took much longer than planned.

3.2.11 Small or Medium-Span Suspension Bridges

Small or medium-span suspension bridges are seldom built today, one reason being that cable-stayed alternatives are more economical. In the presence of steep mountainsides and high-quality rock, as is, for example, the case in Norwegian fjords, suspension bridges at moderate spans of 800 metres or so may be economical, because the anchorages in the mountainside are inexpensive. The anchorage can be made by drilling holes in the rock and installing post-tensioning anchor cables.

For medium-span bridges, main cables may be composed of individual locked coil cables, which may be anchored at the tower tops, or pass over these in individual saddles.

Self-anchored suspension bridges are only feasible for moderate spans, say a maximum 600-metre main span. The bridge deck needs to be temporarily supported from the ground, and hangers need to be installed with a controlled simultaneous stressing operation.

3.2.12 New Challenges (Multi-Span Suspension Bridges, Floating Suspension Bridges, New Material)

The development of future suspension bridge designs includes super-long spans, multi-span bridges, bridges with new cable materials, and floating suspension bridges.

Super-long spans may largely be based on scaling existing construction methods. The cable length alone is a challenge, as there may be practical weight limitations to observe. Also, the tower heights will increase with the length and be more prone to vibration issues. The main cables will also be very large if only two, so, in order to allow parallel worksites, the number of cables may double. This may give rise to new challenges, for example, the installation of cable clamps.

Multi-span bridges may require certain degrees of slip resistance of the main cable in saddles on the central tower(s), which raises additional challenges to the overall bridge geometry configuration and construction of those saddles. Spatial cable systems, i.e., cables that also incline perpendicularly to the bridge axis, will require the main cables to rotate out from their vertical plan during construction, and thus new challenges the contractor to consider and implement solutions.

Suspension bridges with carbon fibre cables have been studied for many years, but the construction of major suspension bridges with such cables is not imminent. When the design matures, this will certainly call for contractors to revisit construction methods.

Construction of suspension bridges on water, i.e., with the towers placed on floating structures, very likely stabilised by a tension leg platform (TLP) system, as known from oil and gas off-shore platforms, are now under study (see section 3.4.3). This will obviously lead to a further step in the complexity of constructing a suspension bridge. One objective is to be able to work within accurate geometric tolerances, while the bridge shifts position due to the change of wind and water conditions. The use of top cables between the tops of floating towers is likely required for the in-service design, but also beneficial for the construction.

3.2.13 Lessons Learnt and Recommendations

The modern way to construct suspension bridges is to build upon construction principles known since nearly the very first suspension bridge, which are, however, over time becoming increasingly refined, reliable, precise, and fast (time-saving). To some extent construction today is business as usual with no or very few significant problems.

Adverse weather conditions occurring repeatedly during construction are a focus point as they may influence the construction time, and in turn, the project economy. Temporary structures are subjected to the harsh environment over long construction periods. There are examples of temporary equipment that has been overloaded to failure, for example by a storm. Suspension bridges are composed of cables which are subjected to vibration, as are the various temporary cable structures, and this vibration may lead to the degradation and potential failure of elements. In many cases, vibration issues during construction have been prevented or mitigated because they were well prepared, whereas in other cases, unexpected and significant vibration occurred. Galloping of cables for tower pull-back systems and partly erected main cables required particular attention and immediate mitigation measures. It is recommended that thorough plans be laid for the various vibration issues that might occur during construction, so that mitigation measures are readily at hand in case vibration actually occurs.

For the construction of anchorages, sometimes below the water line, ingress of water during construction is a difficulty. Also, water seeping through rock can be a reason for underlying delays.

Whereas parts of suspension bridges may be designed with dehumidification systems, during construction the conditions are different. There are examples of various material defects occurring, such as white rust forming on main cable wires during storage, hydrogen embrittlement occurring in bolts subjected to water, etc.

There are examples of self-anchored suspension bridges that proved to be more difficult to construct than first expected.

References

[1] N.J. Gimsing, C.T. Georgakis, Cable Supported Bridges, Wiley, 2021

[2] AASHTO. Guide Specifications for Wind Loads on Bridges During Construction. 1st Edition 2017

[3] David McCullough, The Great Bridge: The Epic Story of the Building of the Brooklyn Bridge, Simon & Schuster, 2007

[4] N.J. Gimsing, East Bridge. A/S Storebæltsforbindelsen, 1998

[5] Magdaléna Sobotková, Hålogaland Bridge Norway. e-mosty, issue 02, 2017

[6] Steel Bridge Erection Guide Specification S10.1-2019, AASHTO/NSBA Steel Bridge Collaboration

https://doi.org/10.2749/sp001.ch03.3

3.3 Arch Bridges

António Adão da Fonseca
Portugal

This subchapter comprises a very summarised overview of arch erection methods, including arch construction with scaffolding (for different scaffolding types,) the cable-stayed cantilever method and the truss cantilever method (with temporary diagonals), rotation or lowering method, and an extremely short reference to other methods, all of them described in detail in extensive available literature [1–5].

Fig. 3.15 Tanzania – SAKIRP Programme by ENABEL

3.3.1 Construction with Scaffolding

In both Eastern and Western cultures, stone or brick arch bridges were built over a wooden formwork supported by a timber scaffolding. This method is still used nowadays (Fig. 3.15)

With the advent of structural iron and steel in the 19th century, iron or steel arches were built using the advancing cantilever method (Fig. 3.16), with bars or small segments transported and placed in position even before electrical cranes were available.

Construction methods used in stone or brick arches or for iron or steel arches were 'imported' into the construction of concrete arches, with the formwork supported by a timber scaffolding, sometimes a work of art in itself (Fig. 3.17) or by a previously built steel arch (Fig. 3.18, Fig. 3.19, Fig. 3.20).

Fig. 3.16 Maria Pia Bridge – Portugal 160 m span, designed by Théophile Seyrig and built in 1877 by Gustave Eiffel

Fig. 3.17 Salginatobel Bridge – Switzerland, 90 m span, designed by Robert Maillart and built in 1930

Fig. 3.18 Albert Louppe Bridge – France (3 spans of 186 m, built in 1930)

The 'navigating' scaffolding conceived by Freyssinet for the construction of the Albert Louppe Bridge (Fig. 3.18) deserves to be mentioned, for it is a masterpiece in itself.

Later, in 1952, Freyssinet came up with another innovation (Fig. 3.19). A pair of cable-stayed scaffoldings advanced from both sides until they were 80 metres apart. Then, a scaffolding segment for the remaining 80-metre span was lifted into place.

Very cleverly, the load capacity of the lateral partial arches was increased by concreting their lower slabs before the lifting of the central arch segment of the scaffolding.

The Arrábida Bridge (Fig. 3.20) was built a few years later, in 1963, with a steel scaffolding that was first used to build one arch and then moved sideways to build the twin arch. That steel scaffolding was planned to be used in the construction of a second bridge upstream in the River Douro, but, sadly, it was left to decay along the riverbanks.

The high cost of arch scaffolding was always a shortcoming in the construction of concrete arch bridges. Entire scaffolding is required because concrete arches must be complete in order to sustain themselves and provide support for the construction of the deck (Fig. 3.21).

Fig. 3.19 Central scaffolding in the three Guaira Viaducts – Venezuela (136 m to 152 m spans)

An interesting construction method had been advanced well before, back in the 19th century, by Joseph Melan, where the formwork was supported from inside by an iron lattice arch (it may be said to be an inside scaffolding) which became the reinforcement of the finished concrete arch. The first bridge built with this method was the Schwimmschule Bridge, in Steyr, Austria, and many others followed with steel lattices (Fig. 3.22).

The Melan method requires an amount of steel for the scaffolding that is much higher than the reinforcement needed for the final concrete arch. However, the merits of this classical construction method explain its current use in record-breaking arch bridges in China, such as the Wanxian Bridge (Fig. 3.23) over the Yangtze River and the Beipanjiang Bridge (Fig. 3.24), the latter presently being the World Record concrete arch [2].

The 'Melan' method was frequently also used to form arch action as early as possible during construction in earthquake-prone countries, like Japan. In the case of Kashirajima Bridge, the 'Melan steel elements' were arranged inside the arch box section (Fig. 3.25).

3.3.2 Cable-stayed Cantilever Method and Truss Cantilever Method (with Temporary Diagonals)

The cable-stayed cantilever launching of the concrete arch alone is quite common nowadays. Two outstanding concrete arch bridges for the high-speed railway over the Alcântara Reservoir in Spain were recently built using this method.

Fig. 3.20 Arrábida Bridge – Portugal (270 m span, designed by Edgar Cardoso)

Fig. 3.21 Arrábida Bridge – Portugal

Fig. 3.22 Echelsbach Bridge over River Ammer – Germany (HaTe photo) 130 m span, designed by Heinrich Spangenberg and built in 1929

Fig. 3.23 Wanxian Bridge – China. 420 m span, built in 1997

Fig. 3.24 High-speed train Beipanjiang Bridge – China. 445 m span, built in 2016

Fig. 3.25 Kashirajima Bridge in Okayama – Japan. 218m span, built in 2004

The bridge designed by the Carlos Fernandez Casado office, led by Javier Manterola Armisen, flies over the River Tagus with a 324-metre span arch rising 70 metres (Fig. 3.26), built with concrete grade C70 [3].

The cable-stayed cantilever launching of the arch was implemented with long backstays anchored directly in the ground, followed by the construction of the columns supported by the arch and by the deck over the arch with two gantries travelling from each side of the bridge, to guarantee loading symmetry.

The other bridge was designed by the Arenas & Asociados office under the leadership of Juan Arenas de Pablo, with an arch also rising 70 metres but spanning 384 metres over the River Almonte (Fig. 3.27). The high-strength concrete C80 helped to achieve a very elegant, curved arch bridge [4]. Again, the arch was built with the cable-stayed cantilever method and the deck by travelling gantries.

The advantages of the use of stay cables for the cantilever launching of arches are clear, but the method of simultaneous construction of arch, deck, and columns together with provisional diagonals to materialize a cantilever truss, is becoming quite competitive. This last method was first used in 1974, in Japan, in the construction of the Hokawatsu Bridge, with an arch spanning 170 metres (Fig. 3.28). The concrete deck was used as a tension member and, therefore, external temporary tendons were installed on the upper side of the deck. Ground anchors balanced deck tension forces.

Just five years later, Ilija Stojadinovic, in cooperation with Vukan Njagulj and Bojan Možina, designed the outstanding Krk Bridges (Fig. 3.29), in Croatia. The arch of the East Bridge spans 390 metres and it is still today a European record.

In fact, the tension chord of the advancing trusses of the Krk East Bridge was not provided by the deck but by cables, thus reducing the weight of the cantilever trusses (Fig. 3.30).

Fig. 3.26 Bridge over River Tagus, at Alcântara Reservoir (finished in 2015). Construction sequence and rendering of the final bridge

Fig. 3.27 Bridge over River Almonte, at Alcântara Reservoir (finished in 2016). Construction sequence and rendering of the final bridge

Fig. 3.28 Hokawatsu Bridge – Japan. 170 m span, built in 1974

Fig. 3.29 Krk Bridges – Croatia

Fig. 3.30 Krk East Bridge under construction

Both the method of erection by cable-stayed cantilever launching, with or without provisional towers, and the truss method, generate increasing compression in foundations and in the advancing half-arches. Elastic deformation, shrinkage, and creep are thus significantly compensated during construction. The cable-stayed cantilever launching is better for geometric control but compression is introduced only partially. The truss method is not as efficient in geometric control but compression is introduced almost entirely, unless the deck is built afterwards, like in the Krk Bridge option. However, if the deck is built simultaneously and is the tension chord of the advancing trusses, extra prestressing is needed in the deck.

In all these construction methods, a quite stiff arch is required to ensure instability does not occur at any stage of the construction.

Robert Maillart designed beautiful slender arches (Fig. 3.31) in the first half of the 20th century.

Later, in the second half of the same century, Christian Menn also designed some outstanding slender arches (Fig. 3.32).

Obviously, a slender arch is only capable of resisting small bending moments. Therefore, a slender arch under live loads requires 'outside' help to ensure that generated bending moments are kept small. A stiff deck accomplishes this, as it is capable of distributing loads 'cleverly' over the various columns. This type of bridge is called a deck-stiffened reinforced concrete arch bridge.

The construction of slender arches has always been very audacious. Robert Maillart and Christian Menn always used complete scaffolding which was taken away only after the bridge was built and the arch/deck 'pair' could interact. The cable-stayed cantilever launching is not appropriate for very slender arches because the required stiffness of the system is difficult to ensure with deformable stays under temperature variation. On the other hand, the 'truss method' has already proved to be a suitable solution. Because the deck is the 'outside' stabilising element of the arch, construction of the arch cannot go ahead of the deck. The first time this method was deployed was in the construction of the Akatanigawa Bridge for high-speed trains, in Japan, with the arch spanning 100 metres and rising 19 metres. The stiff deck and thin arch were built at the same time using a special form traveller (Fig. 3.33).

In the present century, A. Adão da Fonseca invited J. A. Fernández-Ordoñez and Francisco Millanes to join him in the design team for a bridge over the Douro River, between Porto and Gaia, in Portugal. This bridge is a world record for its thin and shallow arch with a span of 280 metres (Fig. 3.34, Fig. 3.35) [5].

Deck construction ahead of arch construction can be seen in Fig. 3.35, as well as the triangulation with temporary diagonal cables.

3.3.3 Rotation or Lowering Method

Reduction in the construction cost of concrete arch bridges is achieved either by making the construction of the scaffolding cheaper or by the elimination of the entire scaffolding. In 1943, the wooden scaffolding for the construction of the second Longeray Bridge over the River Rhône was put in place by the rotation of the scaffolding semi-arches first built almost vertically over the abutments. The same technique was adopted in the construction of the Saboya Bridge over the same River Rhône, but it was Riccardo Morandi who first did it with concrete semi-arches to build the Lussia Footbridge (Fig. 3.36). One year later, in 1954, Morandi adopted a similar procedure for the Paul Sauer Bridge in South Africa (Fig. 3.37), but in this case, without

Fig. 3.31 Schwandbach Bridge – Switzerland (37 m span, built in 1933)

Fig. 3.32 Cascella Bridge – Switzerland (96 m span, built in 1968)

Fig. 3.33 Akatanigawa Bridge – Japan (100m span, built in 1979)

Fig. 3.34 Aerial view of Maria Pia, Infant Dom Henrique (built in 2003) and Luiz I Bridges

Fig. 3.35 Infant Dom Henrique Bridge construction

Fig. 3.36 Lussia Footbridge – Italy (70 m span, built in 1953)

Fig. 3.37 Paul Sauer Bridge under construction

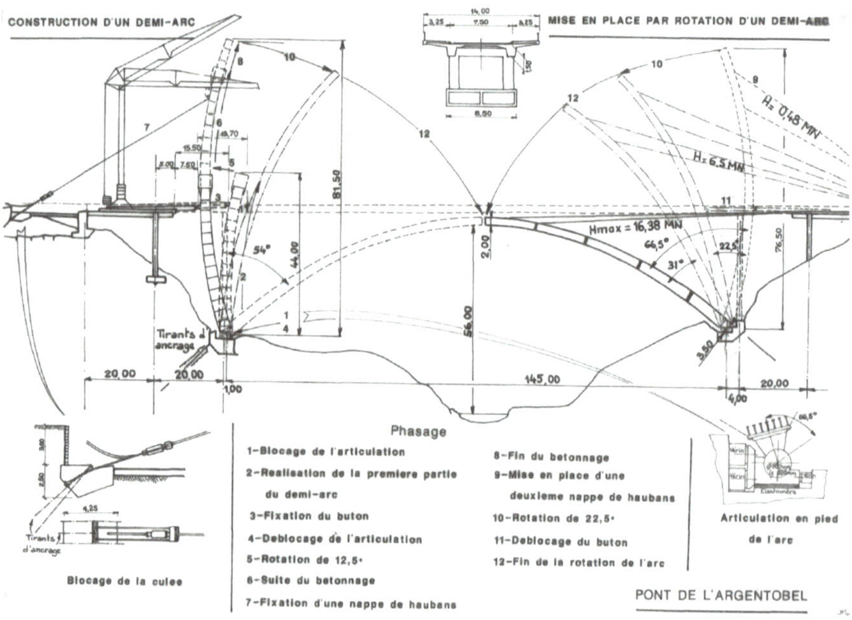

Fig. 3.38 Argentobel Bridge – Germany (145 m span, built in 1987)

the provisional central tower used in the construction of Lussia Footbridge.

In 1987, the same technique was used in Germany to build the 145-metre span concrete arch of Argentobel Bridge (Fig. 3.38).

This 'rotational method' was applied with several ingenious assembly procedures in the construction of the Bridge Arcos de Alconétar (Fig. 3.39), spanning 220 metres over the River Tagus, at Alcântara Reservoir. The two longitudinal beams of the deck, columns supported by the arch, and the arch itself are made of weathering steel.

In 1977, a new technology of horizontal rotation was used in China for the construction of the 70-metre span arch of Suining Jianshe Bridge in Sichuan (Fig. 3.40). Since then, the horizontal rotation method has been applied to build reinforced concrete arch bridges up to a span of 200 metres.

The arch bridge that currently has the largest rotating weight is the Suining Jianshe arch bridge (Fig. 3.41) with a span arrangement of 88.8+160+88.8 metres. The weight of the rotating structure reached 16,800 tonnes. This arch has a parabolic shape and a rise-span ratio of 1/6. The arch was cast in place on a scaffold and then horizontally rotated 26° to cross the busy Shanghai-Hangzhou Expressway to its final position.

Fig. 3.39 Arcos de Alconétar Bridge – Spain. Designed by J A Llombart, J Revoltós and S Wőrner, and built in 2006

Fig. 3.40 Suining Jianshe Bridge – China (70 m span, built in 1977)

Fig. 3.41 Shanghai-Hangzhou high-speed railway Suining Jianshe Bridge – China (160 m span, built in 2010)

3.3.4 Other Methods

Other arch methods are known: for example, the concrete-filled steel tube bridges (CFST) or the double skin tubular arch (DSTA) bridge system. These methods comprise composite solutions and have their own merits in terms of structural response and in terms of the use of materials. The construction methods comprise hybrid combinations of the already presented construction methods.

3.3.5 Lessons Learnt and Recommendations

In light of the various methods for arch erection discussed in this section, it is crucial to consider lessons learnt and recommendations detailed in Sections 2.4.2.3, 2.4.2.4 and 3.1. These sections cover essential recommendations for cantilever construction, incremental launching, and cable-stayed erection techniques.

Additionally, the following key lessons should be considered:

- In arch construction, it is strongly recommended to monitor the displacements and rotations of the arch foundations from the start of erection
- When erecting arches by rotation, it is important to follow all weather condition specifications for operations and to include a specific operational risk analysis. This analysis should consider the duration of operations and potential interruptions. Additionally, dynamic effects such as wind action and the impact of starting or stopping lifts should be addressed
- In cases where arch erection involves the use of provisional cables and the construction sequence requires cycles of re-tensioning and tension reduction operations, it is necessary to consider the potential reduction in the effective cross-sectional area of the cables due to wedge penetration.

References

[1] AASHTO LRFD Bridge Construction Specifications, 4th Edition, with 2020 Interim Revisions

[2] YONG XU, LIE CHEN, HAIQING XIE & JINGTAO HU, Design of the Beipanjiang Bridge on the high-speed railway between Shanghai and Kunming, The Third Chinese – Croatian Joint Colloquium on Sustainable Arch Bridges, Zagreb 2011, pp. 71–82.

[3] MANTEROLA J., MARTÍNEZ A., MARTÍN B., NAVARRO J.A., GIL M.A., FUENTE S. and BLANCO L., Long railway viaducts with special spans: Part 1. Arch construction by balanced cantilever with auxiliary cables, Multi-Span Large Bridges – P. Pacheco & F. Magalhães (Eds.) Taylor & Francis Group, London, 2015, pp. 197–198.

[4] ARENAS J. J., CAPELLÁN G., BEADE H. and MEANA I., Viaducto sobre el Río Almonte en el Embalse de Alcántara para la Línea Ferroviaria de Alta Velocidad Madrid – Extremadura. Talayuela – Cáceres, V Congreso de ACHE. Barcelona: ACHE Asociación Científico-Técnica de Hormigón Estructural, 2011.

[5] ADAO DA FONSECA A., The Infant Dom Henrique Bridge over the River Douro, at Porto, 5th International Conference on Arch Bridges, Madeira – Portugal, P. Lourenço, D. Oliveira & A. Portela (Eds.) University of Minho Press, 2007, pp. 931–960.

3.4 Floating Bridge Construction

Arianna Minoreti | Xu Xiang | Johannes Veie | Simen Hellgren Holtberget | Cato Dørum | Parthasarathi Jena
Norway

3.4.1 State of the Art in Floating Bridges and Future Trends

Floating bridges have been a functional solution since ancient times, especially for temporary crossings, with large applications during the war periods. Several permanent floating bridges were constructed during the last century and concrete has become a possibility as a construction material since 1940. In recent years, other possibilities for floating bridges have been demonstrated to be feasible [1], following the developments of offshore technologies and the experience gained over the last 40 years with tension leg platforms (TLP). Suspension or cable-stayed bridges on tension leg platforms are now an option for long and deep crossings [2]. At the same time, for environmentally exposed areas, the submerged floating tube bridge (SFTB) is a solution that allows naturally reducing the loads on the structure with the submergence. This type of bridge, whose concept was conceived at the end of the 19th century, now represents a feasible solution for the next generation of floating bridges, and some hybrid structures have already been built recently [3].

This section is mainly focused on describing and explaining the essentials of construction methods applicable to floating bridges. For a more in-depth analysis of this type of bridge, an extensive bibliography and regulations are available (among others: [3, 5–7]).

3.4.2 Floating Bridges

3.4.2.1 The State of the Art and the New Challenges

There are several floating bridges today, both straight side-anchored bridges and curved end-anchored bridges. In the first case, the lateral loads are managed using a straight configuration for the bridge deck and anchors connecting the bridge with the seabed along the structure for the entire crossing. In contrast, the second option has a curved alignment of the structure that permits providing anchors only at the shores. The anchoring systems can also be combined, depending on the design needs. Both concepts may require a coupling to a high bridge (such as a cable-stayed bridge) to allow ship traffic to pass the bridge (Fig. 3.42). The ship navigation channel is typically placed towards the bridge abutments.

The concept of straight floating bridges requires anchoring to the sea bottom and expansion joints, which may be quite demanding for long crossings. The load-carrying system is relatively simple, acting as a beam resting on an elastic foundation in

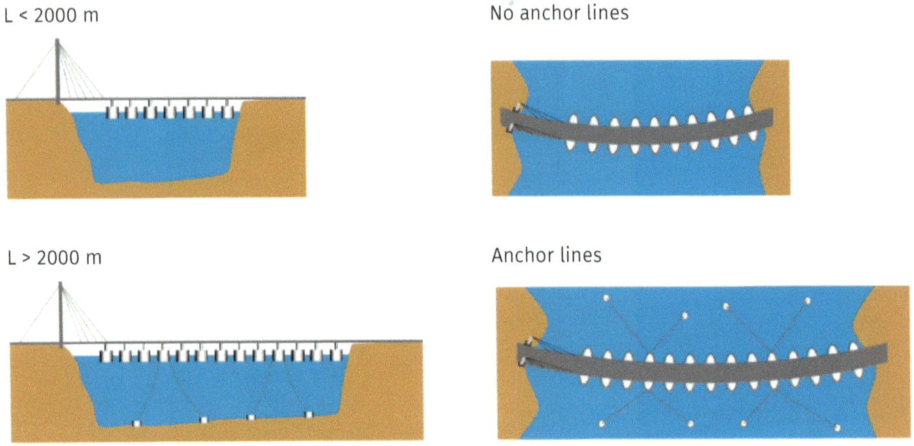

Fig. 3.42 Scheme for crossing with a floating bridge

the shape of discrete or continuous pontoons. Continuous pontoons have been chosen for floating bridges that are crossing sheltered waters. For crossing fjords with more challenging environmental loads, discrete pontoons are used. Both concrete and steel pontoons can be used, but care must be taken to prevent corrosion problems.

The curved floating bridges that exist today are built without any mooring to the seabed, and the overall stability of the system relies on the lateral stiffness given by the curved bridge shape and the fixed connection to the abutments. For longer spans than those built today, it may be necessary to add side-anchoring (mooring lines) to ensure sufficient overall stability.

Engineering News-Record lists the ten longest floating bridges in the world. Among them, the three longest are all located in the Seattle region of the United States of America. The Nordhordland and Bergsøysund floating bridges of Norway are ranked seven and eight for the length of the floating part. The Nordhordland bridge is unique among all floating bridges due to including both a cable-stayed bridge and a floating bridge, where the cable-stayed bridge extended the flexibility of the whole bridge by allowing relatively large ships to pass (Fig 3.43). With a total length greater than 5000 metres, if built, the Bjørnafjord floating bridge (Fig 3.42) will become the longest [8].

3.4.2.2 Construction Stages

Depending on the site of the bridge, different construction techniques can be used. The most promising method is assembling the bridge in a more sheltered environment and towing the structure out to the bridge location. This will be done in a similar fashion as previously done for offshore structures, and planning, facilitation, and training of towing vessel operators is required for the operation. Initial studies show this is a feasible method, but more extensive analyses regarding, for example, fatigue must be done for each site and structure. Parallel to the assembly of the floating part, a high bridge can be constructed on one side of the crossing location, so that the floating part can be connected when the towing commences.

3.4.2.3 Lessons Learnt

The floating bridges in Norway and North America are performing well, both the end-anchored and the side-anchored concepts. The floating bridge is a sensitive structure and the design must be tailored for each crossing. Floating bridges where components are built on an assembly line may be effective with respect to production but can be vulnerable to repetitive errors. Before the bridge is installed, an accuracy control of the geometry must be performed on the structure, to avoid problems in the assembly of the structure or to avoid loads in the structure that have not been evaluated in the design stage. The function test of the pontoon system must be performed before the installation. The construction process must be taken into account during the design, with a particular focus on temporary loads on the structure, necessary marine operations including a temporary stabilizing system, and temporary structures and vessels necessary for positioning and installing the structure.

Fig. 3.43 Phase-5 Bjørnafjord floating bridge concept, end-anchored with side-moorings

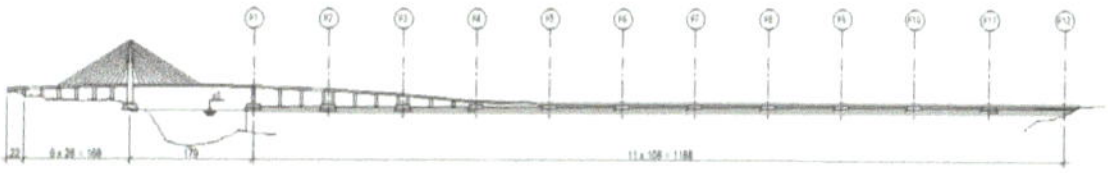

Fig. 3.44 Nordhordland bridge, as built (source: NRPA archive)

3.4.3 Tension Leg Platform Bridges

3.4.3.1 Description of the Structure

The tension leg platform bridge consists of a multi-span suspension bridge whose central pylon(s) rests on floater(s). The floater is inspired by the tension leg platform technology commonly used in the offshore industry in areas of extreme depths. Like tension leg platforms, the floater is anchored to the seabed by means of vertical tethers. The floater can be made of concrete or steel. For bridges with more than two central spans, a top cable has been introduced between the top of the pylons in order to provide the system with longitudinal stiffness to accommodate the asymmetric traffic load situations that would, otherwise, lead to large deflections in the loaded spans [9]. The rest of the superstructure consists of the main suspension cables, pylons, hangers, and an aero-dynamical bridge girder. To be cost competitive the maximum span is limited to the use of a single box girder, typically about 1400 meters long.

3.4.3.2 Construction Stages and Related Problems

The tension leg platform (TLP) typically consists of suction anchors, tethers, and floaters (Fig. 3.45 and 3.46). Both suction anchors and floaters may alternatively be built in steel or concrete. In terms of installation, both alternatives can be done similarly. Suction anchors may be brought to the site on a barge(s) and lifted in place by a semi-submersible crane vessel (SSCV). Floaters may be brought to the site in a wet tow (concrete) or on a submersible barge (steel) and then made self-floating.

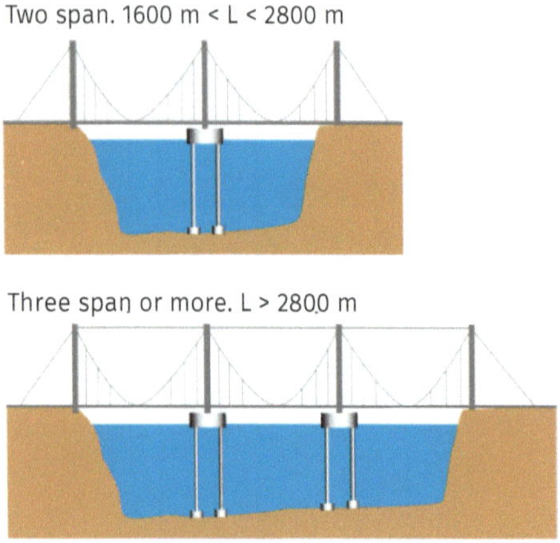

Fig. 3.45 Scheme for crossing with tension leg platform (TLP) bridge

Fig. 3.46 Suspension bridge on tension leg platform, study for the Bjørnafjord crossing in Norway, Statens vegvesen

For the latter case, the process of making the floater self-floating can be done near shore thus allowing for some standby/contingency time before the floater is towed to the site. The tethers may be manufactured in full lengths at a suitable shore location, then launched and brought to the site in a wet tow. At the site, one end will be submerged and connected first to the suction anchors and then to the floater.

Various alternatives for steel pylon construction and installation may be considered. A proven method is a 'stepwise' installation using a climbing crane. This method is conventional (for example, used for the Izmit Bridge) but quite time-consuming as a lot of handling, assembly, and welding must take place at the pylon location. The typical weight of pylon modules that can be handled by this installation method is 50–150 tonnes depending on crane capacity.

An alternative method entails the fabrication of the pylon modules in a workshop and their assembly into a single piece at a quayside location oriented horizontally (Fig. 3.47). The structure is then launched onto a barge, sea-fastened, and towed to site. The pylon is subsequently lifted off the barge, upended in the air, and placed on the floater by means of a large semi-submersible crane vessel. This approach greatly resembles that used for offshore jackets and top-side installation which are conventional operations. A similar sub-optimisation scheme which could be considered is using a medium-sized crane barge to lift and install larger modules in the lower parts either at the shore or on site.

The cable system comprises both temporary and permanent cables.

Temporary Cables:

- Catwalk
- Storm rope system (if required)

Permanent Cables:

- Top cable (if required)
- Main cable
- Hangers.

The catwalk installation for the main cable and to some extent storm rope systems is a standard suspension bridge construction methodology. The hanger installation is also a standard suspension bridge construction methodology.

There is no direct experience with the installation of top cables. The cables resemble those used in cable-stay bridges but the installation will take place using the catwalk where the top cable protection high-density polyethylene (HDPE) pipe will be assembled and rest until the first wires are tensioned. The length of the top cable is typically twice as long as the longest –stay cable. The installation does, however, not appear to be a process with obvious significant challenges aside from getting the catwalk in the right position.

The installation of the top cables in the central spans can take place after establishing the separate catwalk, as follows:

1. Protective HDPE pipes are assembled and welded together from 12-metre-long sections on a series of rollers on the catwalk.
2. Various 7-wire strands are hauled through the HDPE pipe using a hauling line. The number of strands should be sufficient to carry the HDPE pipe and to do an initial stressing.
3. The first multi-strand cables should, as far as possible, be assembled simultaneously in all spans to prevent unacceptable unbalanced forces at the top of the pylons
4. The remaining top cables are erected and stressed in a conventional way.

Before commencing the stressing of the top cables, movements of several metres can be expected of the floating pylons due to winds in the longitudinal direction of the bridge. When the erection and stressing of the top cables commences the movements will gradually diminish and instead, the force in the top cables will vary due to the wind loads.

The top cables in the side span can be installed using the normal stay cable installation procedure if the bridge girder in the side span has been constructed in advance.

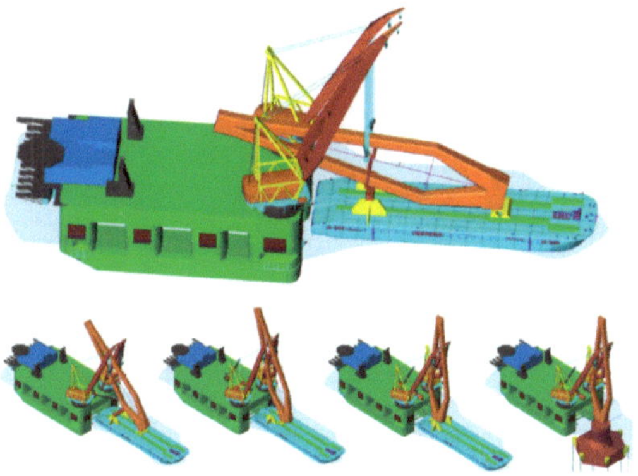

Fig. 3.47 Full pylon installation (Heerema)

The installation of the main cable should be conducted as for normal suspension bridges with one exception. The tension in the cables is typically adjusted based on the measured sag of the cable. Due to the movements of the bridge, this method will not be possible. Instead, precise marking based on measurements of the cables should be used to obtain the same tension in all cables. The main cables are implemented by pulling fully pre-fabricated parallel wire strands (the PPWS method) one by one from one anchor block to the other. This is done by placing the strand on a roller using the pre-installed catwalk.

The bridge deck comprises a closed steel box girder. The bridge deck may be prefabricated at a workshop at a distance from the bridge site in suitable sections and transported by ship to the site. At the bridge site the vessel is positioned under the main cable and the deck section is lifted and attached to the pre-installed hangers. The lifting of deck sections can be carried out by using a heavy lift floating crane or using lifting gantries operating from the main cables.

Welding of the deck sections can start when the deck sections are suspended from the main cable, which has reached its correct geometry by then. During installation, the deck sections are connected to the adjacent sections by temporary hinges or brackets designed to resist environmental loads from wind and temperature.

3.4.4 Submerged floating tube bridge (SFTB)

3.4.4.1 What is a SFTB

The submerged floating tube bridge (SFTB) is a floating bridge, submerged at a defined position under the water level, with water surrounding the tube cross-section. The structure has a closed cross-section like a tunnel but the structural behaviour is that of a bridge. For short crossings, the structure could have a single span between the shores, while for longer crossings [10] the structure can rely on columns (in the case of shallow waters) or be vertically stabilised with floating pontoons or tethers connecting the structure with the seabed (Fig 3.47 and 3.48).

Design and construction are strongly influenced, depending on the loads on the structure, by the local characteristics of the installation area and the available facilities onsite.

3.4.4.2 Construction and Installation

The structure is a combination of known technologies [11]: the tube structure is similar to that of subsea tunnels, the tethers are the same as those installed for offshore oil production facilities on tension leg platforms, and the pontoons are the same as those used in floating bridges. Therefore, many construction operations are similar to those related to these structures.

The SFTB can be constructed mainly following three different schemes:

1. The elements can be fabricated, transported, and installed to the final position one by one.
2. The structure could be produced using the incremental launching method, from one of the shores.
3. The structure could be fabricated, assembled into a complete configuration, and then towed to the final position for installation.

The fabrication of the single elements is done in dry docks. After the element or the group of elements is completed, the dock is flooded and all the measurements and testing can be performed. The testing of the elements before installation, with the necessary measurements, for example, of weight and horizontal centre of gravity, is of fundamental importance due to the crucial role of the weight in a structure that relies on its self-weight and on the permanent and variable ballast to counterbalance the buoyancy. This operation would be more difficult through incremental launching, and the use of temporary cables or pontoons (or buoyancy towers) is necessary for the control of the elements.

The joints between the elements must be made according to the design: if the structure is designed as a continuous beam, the joints must provide the necessary stiffness. This can be reached, for example, by using joints similar to the ones of the subsea tunnels for the temporary phase, while the joints are sealed and the structure is completed with the use of prestressing cables (Fig 3.49, Fig 3.50). During the subsea joining of the elements, a guiding system is necessary for connecting the new element using several anchor bolts, and cofferdams are used for dewatering the space inside the SFTB. The final concreting and pre-stressing operations will be possible using temporary shafts.

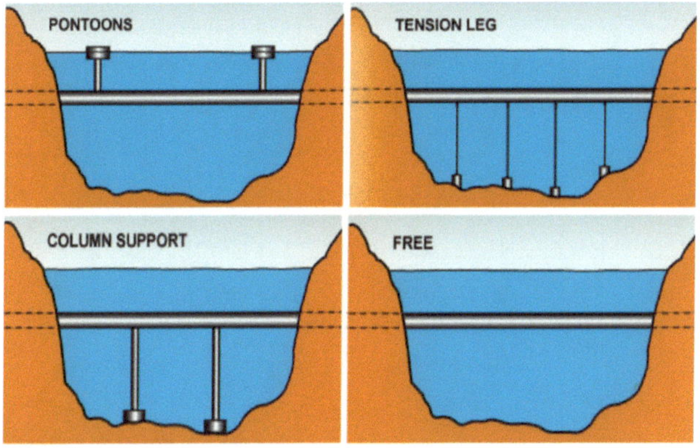

Fig. 3.48 ▶ Scheme for possible crossings, A.W.Solerød – Statens vegvesen

Fig. 3.49 ▶ Submerged Floating Tube Bridge, Study for the Sognefjord crossing in Norway, Statens vegvesen

When the structure is in its final installation position, the vertical stabilising elements, if any, are connected: floating pontoons, tethers, or columns. While tethers and columns must already have been installed when the tube structure is in place, the pontoons can be transported and connected one by one. This operation implies the use of temporary towers to control the structure during the submergence operations.

The SFTB could have landfalls and a support caisson, or support boxes, at the shores to manage the transition between the bridge and the rock tunnel onshore. For those elements, consistent work to prepare the seabed is required before the elements can be installed. This operation is similar to that required for an immersed tunnel.

Measurements are taken during construction, joining, and installation phases, to always keep tolerances under control.

3.4.4.3 Challenges

Between the structure's construction and installation, there is a phase of towing the elements or the structure to the final position. This operation would be more complicated for longer and slender structures and all the forces acting on the elements during the transportation to the final site must be considered in the design. There is limited experience in the towing procedure of long and slender structures, therefore, this operation must be designed in detail taking as a reference the experience gained in the towing procedures of the offshore

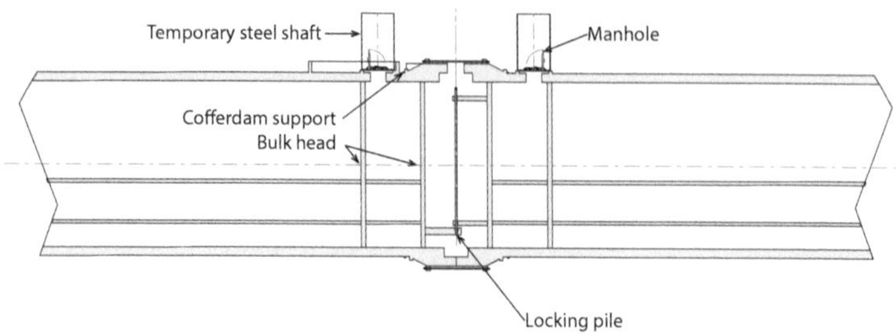

Fig. 3.50 SFTB initial joining of elements, Statens vegvesen

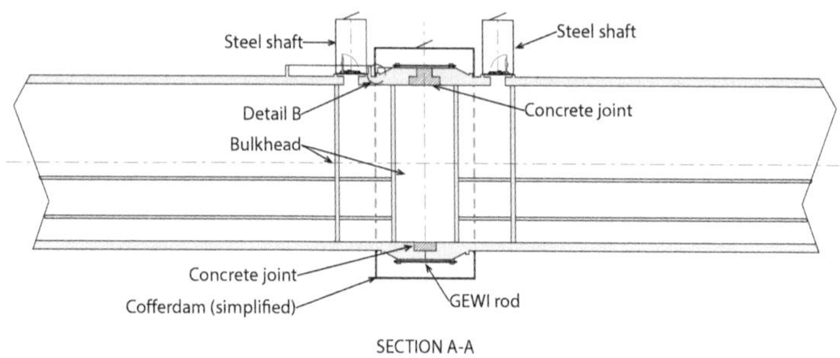

Fig. 3.51 SFTB assembly, Statens vegvesen

platforms, which are, nevertheless, more rigid and punctual structures compared with an SFTB.

3.4.5 Lessons Learnt and Recommendations

While floating bridges have been built since ancient times, suspension bridges on TLP structures and the submerged floating tube bridge (SFTB) haven't been realised yet, even if, for example, the SFTB was conceived for the first time in the late 1800s. This is mainly due to the lack of technology available in the '80s with the developments in the offshore field. Since then, with the growing maturity of the technology and the related confidence, hybrid solutions have started to be built. The challenges of long and slender structures are very dependent on the construction and installation phases, which need to be planned taking into account the loads and responses occurring during these phases. The possibility of partial or total construction in protected areas, where the environmental loads are less challenging with respect to the final position of the structure, should be evaluated, together with the loading situation derived by the transport of the structure to the final position. The challenges for the coordination of the vessels for the transport of the structure should be addressed with specific studies. The construction of the anchorages should be planned together with the temporary stabilizing system and the planning of the phases for the connection to the structure. Experience for the realization of the moorings could be gathered from the offshore industry, taking into account that the distribution of the elements along the crossing increases the possibility of having different soil characteristics for the anchorages. This implies the need to rely on the geotechnical information gathered for the crossing along the whole crossing length. The necessary time to complete the construction causes the necessity of identifying protected weather windows for the construction and installation to proceed. Marine operations may become extremely complex for large crossings and shall be planned based on recognized standards and recommended practices.

References

[1] Minoretti A, Xiang X, Johansen IL, Eidem M. The Future of the Tunnel Crossing: The Submerged 572 Floating Tube Bridge. Struct Eng Int 2020:1–5. doi:10.1080/10168664.2020.1775165.

[2] Holtberget, S.H., Veie, J., Villoria, B., Jena, P., Dørum, C., Madsen, P. Concept overview of a multi-span suspension bridge on floating Foundations. 39th IABSE Symposium – Engineering the Future September 21–23 2017, Vancouver, Canada

[3] Fib (Fédération Internationale du Béton) Bulletin No. 96. Guidelines for Submerged Floating Tube Bridges. Guide to good practice (ISBN 978-2-88394-144-1, October 2020)

[4] DNV-OS-C105. Structural Design of TLP (LRFD). (2021). DNV.

[5] DNV-OS-C301. Offshore Standard, Stability and Watertight Integrity. (2021). DNV

[6] Norwegian Handbook N400, Bruprosjektering, ISBN: 978-82-7207-680-0

[7] API RP 2T Recommended Practice for Planning, Designing and Constructing Tension Leg Platforms. (2010). American Petroleum Institute.

[8] Xiang, X., Eidem, M., Minoretti, A., Søreide, T., & Nestegård, A. (2017). Global analysis of Submerged Floating Tube Bridge (SFTB): the design case of crossing the Bjørnafjord Norway. Proceedings of the World Tunneling Congress 2017 – Surface Challenges – Underground Solutions. Bergen, Norway: Norwegian Tunnelling Society.

[9] Veie, J., Holtberget, S.H., Jena, P. (2016). Crossing the Bjørnafjord with a multi-span suspension bridge on floating foundations – Concept Overview. IABSE Congress: Challenges in Design and Construction of an Innovative and Sustainable Built Environment, Stockholm, Sweden, 21–23 September 2016

[10] Minoretti, A., Eidem, M., & Aasland, T. (2019). The Submerged Floating Tube Bridge for the Norwegian Fjords. Proceedings of The 29th International Ocean and Polar Engineering Conference, 16–21 June, Honolulu, Hawaii, USA (p. 686). International Society of Offshore and Polar Engineers.

[11] Minoretti, A., Johansen, I., Xiang, X., & Eidem, M. (2019). Proven technology for a new structure: Submerged Floating Tube Bridge. 20th Congress of IABSE 2019 – The Evolving Metropolis (p. 15692). New York City, USA: Curran Associates Inc.

4

Risk, Time, Logistics and Costs

4.1 Construction Method Risks

Hugo Coelho | J.M. Sousa Soares | Pedro Pacheco

Portugal

4.1.1 Acronyms used in this Sub-chapter

In the present sub-chapter, the following acronyms are used:

BDE	Bridge Deck Erection Equipment
MSS	Movable Scaffolding System
EN	European Norm
CC	Consequence class
EXC	Execution class (Eurocode)
γ_G	Permanent load partial coefficient ($\gamma_{G,SUP}$ for unfavourable and $\gamma_{G,INF}$ for favourable)
γ_Q	Live load partial coefficient ($\gamma_{Q,SUP}$ for unfavourable and $\gamma_{Q,INF}$ for favourable)
γ_C	Concrete load partial coefficient
γ_W	Wind load partial coefficient
γ_{ID}	Imposed displacement partial coefficient
ψ_0	Combination coefficient
ψ_2	Combination coefficient
γ_M	partial factors for strength parameters and resistances (Including γ_{M0}, γ_{M1}, γ_{M2}, γ_{M3}, $\gamma_{M3,SERV}$, γ_{M5}, $\gamma_{M6,SERV}$ and γ_{M7})
IAP	Instruction regarding loads for roadway bridge design – "Ministério do Fomento Spain
SLS	Service limit state (for load combinations)
ULS	Ultimate limit state (for load combinations)
ACC	Accidental design situations (for load combinations)
EQU	Static equilibrium (for load combinations)
G	Permanent loads
R	Steel reinforcement weight (load)
C	Concrete load (C1 – 1st stage and C2 – 2nd stage)
W	Wind loading (W1 operational wind, W2 intermediate storm and W3 storm wind)
Q	Live load
LOC	Locomotion loads
SLI	Sliding loads
ID	Imposed displacement
ISO	International Organization for Standardization (Norms)
R(O) or R	Risk of an Occurrence
P(O)	Probability of an Occurrence
C (O)	Consequence of an Occurrence
POG	Probability of Occurrence Generation
PF	Probability of Failure
β	Reliability index
Φ	Cumulative distribution function of the standardized Normal distribution
HSE	Health and Safety Executive
HRA	Human Reliability Assessments
HEART	Human error assessment and reduction technique
EPC	Error Producing Condition
HEP	Human Error Probabilities
BDK	Banedanmark

4.1.2 Overview of Standards and Regulatory Background

In order to be able to particularise, contextualise, and develop content in a more in-depth approach, this section and the following sections will gradually move away from general considerations to focus on a specific case of BDE – Movable Scaffolding Systems (MSS) as an example.

Movable scaffolding systems are used for in-situ construction of concrete bridge decks.

MSS structures are often seen as vulnerable when compared to conventional permanent structures. This apparent vulnerability (founded or not) arises from several factors and their combinations, some of the most important being: MSS are mobile structures designed and dimensioned to be lightweight but there is little project expertise for this type of structure. MSS structures often incorporate hydraulic components, they are usually isostatic structures for certain failure configurations, and these structures are subject to successive assembly and dismantling. MSS are often framed as temporary structures, and there is no specific regulatory framework for this type of structure. To overcome these challenges, a regulatory framework and an approach for numerical risk analysis which expands to other existing studies [1, 2] to complement the calculation is proposed.

4.1.2.1 Proposed Regulatory Framework for MSS Design

This section will look at the regulatory framework for the structural design of MSS, and put forward a proposed approach. The aim here is not to impose criteria but to share experience, and propose a background for discussion or establish a starting point for those who are not familiar with this topic.

The application of existing codes to the design of MSS structures is complex and challenging. The standards issued by the United States (US) and in the European context simply do not cover all aspects related to MSS design. However, it appears that the design framework implemented in the Eurocodes is more robust and comprehensive than that specified in the US codes. Alternative design approaches based on a risk management framework may provide a more comprehensive and thorough analysis.

As a reference, the authors recommend adopting the following regulatory framework which can be adjusted on a case-by-case basis. Table 4.1 refers to the proposed codes and Table 4.2 refers to the proposed parameters and coefficients.

Table 4.1 MSS Design and Execution – proposed Codes

Code/Source	Relevant Contents
EN 1990 [3]	General arrangement for load combinations; Definition of Consequence Class.
EN 1991-1-4 [4]	Wind load evaluation.
EN 1991-1-6 [5]	Definition of concrete load coefficient for the design of local elements. Assessment of wind return period.

Code/Source	Relevant Contents	
IAP – Instruction regarding loads for roadway bridge design – *Ministério do Fomento* [6]	Complementary information to EN 1991, regarding wind assessment. May be useful as a complement to the Spanish Manual [7].	
EN 1993 (Eurocode 3)	**EN 1993-1-1** [8]	Material properties for design; Guidance on second-order analysis; Section analysis formulation; Buckling analysis (including lateral buckling).
	EN 1993-1-5 [9]	Inferior chord local analysis (Patch Loading).
	EN 1993-1-8 [10]	Bolted joints; Welded joints; Pinned joints; HSFG joints.
	EN 1993-1-10 [11]	Material properties selection and through-thickness analysis.
	EN 1993-1-11 [12]	Tension bars analysis.
EN 1090 [13]	Technical requirements for execution of steel structures.	
Manual de cimbras autolanzables [2]	Complementary information to European norms, regarding design and manufacture, MSS operational limits, MSS specific loading (whenever values defined in Eurocodes are inadequate).	

Table 4.2 MSS Design and Execution – proposed parameters and coefficients

Parameter	Proposed value	
Consequence Class [3]	CC2	
Execution Class [13]	EXC2 (by default) EXC3 for welding inspection and acceptance criteria (critical welds)	
Fresh reinforced concrete volumetric weight	26 kN/m^3	
Q – Live Load [2]	Concrete Pouring Stage	0.75 kN/m^2 (limited to a total of 60 kN)
	Launching Stage	0.75 kN/m^2 (limited to a total of 20 kN)
Operational wind speed limit [2] – W1	Usually limited to 60 km/h (peak value)	
Moderate storm wind speed – W2	Evaluate on a case-by-case basis*	
Severe storm wind speed [4, 12] – W3	Evaluate on a case-by-case basis. Return period between 10 and 50 years is recommended*	

Parameter	Proposed value	
γ_G – Permanent load partial coefficient [3]	Set A: Equilibrium	$\gamma_{G,SUP}$ = 1.10 (Unfavourable) $\gamma_{G,INF}$ = 0.90 (Favourable)
	Set A: Resistance associated with static equilibrium	$\gamma_{G,SUP}$ = 1.35 (Unfavourable) $\gamma_{G,INF}$ = 1.15 (Favourable)
	Set B: Resistance	$\gamma_{G,SUP}$ = 1.35 (Unfavourable) $\gamma_{G,INF}$ = 1.00 (Favourable)
γ_Q – Live load partial coefficient [3]	Chapter 7 Set A and B Chapter 8 $\gamma_{Q,SUP}$ = 1.50 (Unfavourable) $\gamma_{Q,INF}$ = 0 (Favourable)	
γ_C – Concrete load partial coefficient [2]	1.35	
γ_W – Wind load partial coefficient [3]	1.50	
γ_{ID} – Imposed displacement partial coefficient	1.50	
ψ_0 - Combination coefficient [3]	min 0.7 (live loads) min 0.6 (wind loads)	
ψ_2 - Combination coefficient [3]	min 0.6 (live loads) min 0 (wind loads)	
γ_{M0} [8]	1.0	
γ_{M1} [8]	1.0	
γ_{M2} [8]	1.25	
γ_{M3}(EN 1993-1-8) [10]	1.25	
$\gamma_{M3,SERV}$ [10]	1.0	
γ_{M5} [10]	1.0	
$\gamma_{M6,SERV}$ [10]	1.0	
γ_{M7} [10]	1.0	

In case of the nonexistence or non-applicability of standard regulations, specific design procedures should be adopted and explicitly described in the design assumptions.

* According to the previous table, three levels of wind speed are recommended:

W1 is the operational wind speed limit. This speed corresponds to the operational limit of the MSS, beyond which critical operations are restricted – usually launching and concrete pouring. Normally, the wind speed in these operations is controlled by forecasts and by onsite monitoring using an anemometer. The value usually adopted for W1 is limited to 60 km/h (peak velocity) – but this may vary from one application to another and different values may also be set for different operations. Above this value, the working conditions become very adverse. However, its definition must be based on daily maximum data, and it is reasonable to be less restrictive in the case of construction in very windy areas – given that the limit speed is approved by the MSS designer.

W3 corresponds to storm winds, calculated in accordance with EN 1991-1-4 [4].

When defining the return period, the expected duration of the construction can be taken into account by applying EN 1991-1-6 [5] (which establishes a return period of 10 years for a period of execution of less than 1 year). However, the not infrequent possibility of stoppages during construction should be taken into account when determining the execution period. The storm scenario W3 is considered foreseeable and specific bracing must be provided for these situations. Storm bracings must be easy to assemble – otherwise, the probability of inadequate assembly or even non-assembly is not negligible.

W2 wind corresponds to an intermediate storm scenario to safeguard unpredictable situations, in which the MSS is caught in an unbraced situation. As a first approach, a value between 90 and 120 km/h is recommended and its determination can have a probabilistic basis [14].

The incorporation of wind scenarios in the design serves to cover sudden phenomena, such as the one that occurred at the Pukkelpop festival in Belgium in August 2011, in which a sudden storm (the like of which had never previously been experienced in Belgium, striking the concert field, and whipping up winds of more than 128 km/h) led to the collapse of provisional structures (stages and tents) with a tragic outcome of 5 deaths and more than 100 persons severely injured.

Permanent structures – such as bridges – are already dimensioned for high return periods, but the BDE (which is the main matter of concern here) is often under-dimensioned for wind action, presenting great vulnerability in storm scenarios.

Given the previous remarks on wind speeds, the following limit states must be considered in the design: Service limit state (SLS), Ultimate limit state (ULS), Accidental design situations (ACC), and Static equilibrium (EQU).

In a situation in which the deck section is poured in 2 stages, Table 4.3 presents the typical base combinations that are recommended.

Table 4.3 Typical base combinations

Stage / Scenario	Combination Type	Typical Base Combination	
Concrete pouring preparation for 1st Stage	SLS / ULS	G + R1 + W1	(E1)
Severe storm – concrete pouring preparation for 1st stage	ULS	G + W3 + R1	(E2)
Concrete pouring 1st stage	SLS / ULS	G + C1 + Q + W1	(E3)
Sudden storm – concrete pouring 1st stage	ULS	G + C1 + W2	(E4)
Concrete pouring 2nd stage	SLS / ULS	G + C1 + C2 + Q + W1	(E5)
Severe storm – curing 2nd stage	ULS	G + C1 + C2 + Q + W3	(E6)
Launching	SLS / ULS	G + Q + ID + LOC + W1	(E7)
Launching equilibrium	EQU	G (x 0.9 – stable; x 1.1 – unstable) + 1.5 × Q (unstable) + 1.5 × W1 (unstable)	(E8)
Sliding	SLS / ULS	G + Q + SLI + W1	(E9)

In which,

G – Permanent loads such as structure deadweight;
R – Steel reinforcement weight;
C – Concrete load;
W1 to W3 – Wind loading according to Table 4.2;
Q – Live load;
LOC – Locomotion loads;
SLI – Sliding loads;
ID – Imposed displacement in launching stage, deriving from manufacturing imperfections, errors in positioning of supports, variation of bridge longitudinal slope and potential deformation of supports.

4.1.2.2 Specific Aspects of MSS Design

The tables above provide some guidance on ULS, EQU, and SLS combinations that, although not made specifically for MSS, can be applied to MSS. In addition to these combinations, accidental combinations (ACC) should be foreseen in situations where occurrences that are not unlikely may lead to a reduction in structural safety. Among others, and as an example, the following situations can be framed in accidental scenarios:

- Failure during assembly of formwork suspension bar – insufficient tightening of the coupler (Overhead MSS);
- Damage to the thread of a support frame anchor bar during frame assembly (Overhead MSS);
- Insufficient tensioning of the pier brackets prestressing bars (Underslung MSS).

In general terms, the Spanish manual states that the execution of a manoeuvre forbidden by the operations manual may give rise to an accidental situation [7].

Accidental combinations must be analysed case by case, according to the characteristics of the equipment and the construction process. The risk analysis discussed later in this chapter can provide a methodology for the identification of the most relevant accidental situations. It should be noted that, in certain cases, accidental scenarios cannot be solved with simple dimensioning exercises and may require the introduction of redundancy or even changes to the concept of structural systems or subsystems to attain an appropriate risk level.

Other aspects such as fatigue [15, 16], seismic events [17] and temperature effects [7] should be considered.

4.1.3 Proposal of a Numerical Approach to Risk Analysis

4.1.3.1 Fundamentals of Risk Analysis

When applied to the design of a structure with the characteristics of a movable scaffolding system, risk analysis is a multifaceted tool that permits the integration of the different aspects, systems, or subsystems of the equipment that are relevant to its safety in the operational phase – including the human factor. The essential point of the integration of these different aspects is the adoption of a common language, allowing the risk associated with the failure of a mechanical or structural component or with a human mistake to be translated into the same reference system.

Structural design standard tools only provide a limited measure of safety, based on partial conclusions (developed for current structural elements). Due to the particular nature of MSS structures, an integral analysis is required for proper risk assessment.

Risk analysis is proposed in the machine directive [18] and guidance for its implementation is given in ISO 12100 [19]. This is a very valuable tool to assist and complement the structural analysis, particularly for MSS structures. In temporary structures, unlike permanent ones such as buildings or bridges, operation is a critical aspect of risk analysis.

Risk analysis focuses on the characterisation of different events or scenarios with a probability-consequence binary.

Naturally, the ideal would be to have a clean database with a sample size large enough to allow assigning probabilities on a statistical basis – an example is shown further on. Unfortunately, this is not the most usual situation – but that does not rule out the application of risk analysis.

In cases where there is no statistical basis, the design should assign the agreed probability values and consequences following a common benchmark, which should establish the risk acceptance limit.

This is the approach that will be developed in this chapter. The established probability and consequence values have been used in different MSS applications and, as such, have undergone some fine-tuning over the years. Nevertheless, all suggested values should always be analysed with a critical perspective.

Should it prove challenging to converge on a benchmark, there is always the possibility of submitting the analysis to an independent review or of drawing up a parallel analysis.

Getting into the language of risk analysis begins with a few basic definitions:

- Risk factor: Every factor of physical, environmental, or human nature with the potential to disturb a process, causing an occurrence or harm
- Hazard occurrence: Disturbance in a process
- Extreme scenario: Incident or accident that is the result of a sequence of occurrences
- Maximum acceptable risk: by convention, this is the unitary risk.

An occurrence takes place when a process is disturbed (caused by a certain risk factor). The nature of the risk factor may be physical, environmental, or human-based.

The risk associated with a given occurrence [R(O)] is defined as the product between the probability [P(O)] and the consequence [C(O)] of this occurrence [20].

E11 $$R(O) = P(O) \times C(O)$$

In order to evaluate risk, a standard measure is required – a benchmark needs to be established. Thus, the following occurrence may be established as a reference:

'Collapse of a structural element, with normal complexity, designed by an autonomous designer under a normal pressure context, and whose design was based on reglementary Ultimate Limit State (ULS) criteria.'

By convention, the risk associated with this occurrence is established to be unitary (R = 1.0) and constitutes the maximum acceptable value. This is so because the risk associated with this occurrence is commonly accepted and, moreover, the safety factors used in structural analysis by the ULS approach are backed up by reliability assessments [3]. The unitary risk is the willingly taken risk.

A single risk factor may give rise to one or more than one occurrences and each occurrence is characterised by a pair of probability and consequence values.

When starting from a basis of insufficient or poor-quality statistical data, empirically-based (conventional) uniform criteria may be established for the systematic assignment of probability of occurrence generation (POG) values to different risk factors (Table 4.4) and for the definition of a consequence scale (Table 4.10). Values are assigned assuming that a set of occurrences deriving from the same risk factor is characterised by a sole probability value.

4.1.3.2 Probability of Occurrence Generation

4.1.3.2.1 General Context

The first element of the risk analysis is the probability of occurrence. The reference values to be considered are summarised in Table 4.4.

The items in the first line of Table 4.4 – which is the reference line – underpin the reliability index β [3] presented in Table 4.5. If the general assumptions for design and fabrication recommended in this chapter are implemented, this should allow for attaining a reliability index of 3.8 for current structural elements under a scenario of a reference period of 50 years for new structures.

Looking at the definition of a reliability index pursuant to EN 1990 [3]:

E12 $$P_F = \Phi(-\beta)$$

Where:

P_F is the probability of failure

β = the reliability index

Φ is the cumulative distribution function of the standard normal distribution.

The relationship between Φ and PF is given in Table 4.5.

Table 4.5 shows that the targeted reliability index of 3.8 (close to β =3.72) corresponds to an approximate probability of failure of 0.0001 ($P_F = 10^{-4}$). Hence, the probability of occurrence generation (POG) in Table 4.5 seems to be reasonably coherent with the reliability rationale.

4.1.3.2.2 Correction Probabilities

Going back to the first line of Table 4.4, it is intuitive that the probability of failure of a structural element depends largely on the profile of the designer and even on the context in which the project is developed – a context that can be of greater or lesser pressure.

The structural analysis of a movable scaffolding system is of a significant degree of complexity. Hypothetically, if the structural design was entirely developed by a trainee without supervision, the probability of failure would be clearly higher than that presented in the tables above and the risk of structural failure would be prohibitive.

This type of consideration can and should be included in the risk analysis. Table 4.6 presents a proposal for correcting the probability of structural failure according to the profile of the designer and the context of the project.

The same type of reasoning can be applied to human failures in operational tasks.

An experienced operator whose action is properly supervised has an intuitively lower probability of failure than an inexperienced, unsupervised operator.

Human error has a considerable probability of occurring. Indeed, it has been found that the operator's lack of experience, ignorance, and failure to conduct checking are some of the main accident causal factors, according to a study con-

Table 4.4 Summary of probability attribution criteria

POG	Probability	Type of Risk Factors	Examples of Risk Factors
0.0001	Neglectable	Current structural elements	Girder chords, current bolted and welded connections
0.001	Very Small	Structural elements specially conditioned by manufacturing or assembly operations	Friction supports
		Equipment with very high reliability (hydraulic element)	Hydraulic jacks Pressure relief valves
0.01	Small	Equipment with moderate reliability or robustness (electric element)	Electric wired anemometer
0.1	Considerable	Equipment with low reliability or low robustness	External digital pressure gauges
		Human error in simple tasks in which the risk of failure is intuitive	Consideration of wind speed prior to MSS launching Tightening of suspension threaded bar couplers
1	High	Human error in routine tasks Certain event	Consideration of luminous tower lamps Completion of a checklist

* POG: Probability of Occurrence Generation

Table 4.5 Relationship between β and P_F [3]

P_F	10^{-1}	10^{-2}	10^{-3}	10^{-4}	10^{-5}	10^{-6}	10^{-7}
β	1.28	2.32	3,09	3.72	4.27	4.75	5.20

Table 4.6 Probability of Component Structural Failure – impact of designer profile and project context

	Normal Pressure (normal dates, normal complexity, no overload)				High Pressure (demanding dates, high complexity, design overload)			
	Without Revision	Internal Supervision	Internal Revision	External Revision	Without Revision	Internal Supervision	Internal Revision	External Revision
Inhouse autonomous designer	***0.0001***	***0.0001***	***0.0001***	***0.0001***	0.001	***0.0001***	***0.0001***	***0.0001***
Inhouse senior designer	0.001	***0.0001***	***0.0001***	0.001	0.01	***0.0001***	***0.0001***	0.01
Inhouse junior Designer	0.1	0.001	***0.0001***	0.001	1	0.1	***0.0001***	0.01
Outsourced autonomous designer – Validated Profile	0.001	***0.0001***	***0.0001***	0.001	0.01	***0.0001***	***0.0001***	0.001
Outsourced senior designer – Validated Profile	0.01	0.001	***0.0001***	0.01	1	0.01	***0.0001***	0.01

Table 4.7 Impact of operator skill and supervision in human-caused failure probability

	Probability of Failure	
Type of complementary action	**With qualified / experienced supervisor**	**Without qualified / experienced supervisor**
Checks with records obligation (e.g. measures, photos)	**[0.1]**	0.1 – **[1.0]**
Redundant checks: Operator + Supervisor	**[0.1]**	**[1.0]**
Validations approved by an external entity (double validation)	**[0.01]**	**[0.1]**
Operation checklists – before the operation starts	**[0.1]**	**[1.0]**
Checks during operation	0.1 – **[1.0]**	**[1.0]**
General checks or without well-defined acceptance criterion	**[1.0]**	**[1.0]**
Operations with checklists where there is a clear perception of risk	0.001 – [**0.01]**	0.01 – **[0.1]**

Note: Recommended values in brackets [].

ducted by the Health and Safety Executive (HSE) in the United Kingdom [20]. Measures to reduce the probability of a certain hazardous event occurring are proposed in Table 4.7, accounting for the experience/qualification of the operator and the existence or absence of supervision.

There are techniques to perform human reliability assessments (HRA) that take into account the existence of risk exponentiation factors. The human error assessment and reduction technique (HEART) is based on the principle that every time a task is performed there is a possibility of failure which is affected by one or more error producing conditions (EPC) – for instance: distraction, tiredness, cramped conditions, etc. – to varying degrees. Among the available bibliography, special reference should be made to the study of human error probabilities (HEP) when performing railway operations, commissioned by Banedanmark (BDK) [21]. Table 4.8 presents an overview of generic tasks and resulting human error probability (HEP) estimates and, when available, intervals within the 5–95 % range. There are seven basic tasks, with the first ('task 0') representing the human performance limit.

Although Table 4.8 was prepared for a quite different application, interesting information can be drawn from it:

- The human limit in single tasks is set on HEP = 0.0001 – the minimum probability of occurrence generation (POG) in Table 4.4;
- For tasks performed with urgency and stress – which is usually the case in MSS operation – HEP varies between 1E-1 and 3E-1, the same order of POG in Table 4.4;
- In tasks performed as a team ('task 0') – HEP decreases – this is coherent with the adjustment presented in Table 4.7;
- Table 4.8 was drawn for an activity with trained personnel and estab-

Table 4.8 HEP associated with generic tasks – railway operations (Source [21])

Generic Tasks			
Title	**Description**	**HEP**	**Ref.**
0. Human performance limits	Highest obtainable reliability (minimal error rates).	1E-4 (single) 1E-5 (team)	[22]
1. Simple routine	Simple, familiar, and frequent task, skill-based or rule-based, for which procedures are available	4E-4 (0.00014 – 0.0009)	[23]
2. Nontrivial familiar	Familiar, relatively frequent task, requiring knowledge-based performance and judgment. (Normal tasks involving a deviation from planned operations.)	1.6E-2 (0.012 – 0.028)	[23]
3. Communication, routine	Familiar content routinely conveyed, where at least a limited template for communication is available, and error capture and correction are possible.	6E-3	[24] [25]
4. Communication, nonroutine	Unforeseen, novel content, where no template – or only a rudimentary one – is available.	3E-2	[22]
5. Emergency scenarios – known	Task characterised by some urgency and stress due to safety or production concerns, for which an action plan (template, script) and relevant information are available.	1E-1	[22]
6. Emergency scenarios – unknown	Task characterised by high degree of urgency and stress due to safety or production concerns, where no appropriate action plan (template, script) is available, and relevant information is uncertain or missing.	3E-1	[22]

lished routines – in such situations and if urgency and stress are not a relevant factor, human accuracy increases considerably ('tasks' 1 to 4).

There are techniques with different levels of complexity. Nonetheless, and particularly for those engaged in first approaches to risk analysis, the recommendation is to try to keep it as simple as possible and to be conservative – always playing on the safe side.

To evaluate the risk associated with a certain occurrence, it is necessary to assign a value quantifying the consequence of that occurrence. The consequence table should be built top-down, where the maximum consequence corresponds to an unacceptable outcome. The remaining consequence levels should be established in a gradual and proportional manner.

4.1.3.2.3 Statistics on Equipment Failure or Human Error Probability (HEP)

A long-term monitoring, research, and development project for risk analysis in construction has been undertaken by the authors of this section over nearly 20 years, comprising the evaluation of 21 projects in 11 countries, on four continents.

The method used to identify the typified probabilities is based on a systematic study strictly in line with the following assumptions:

Table 4.9 Occurrence Probability Scale

Objects of Occurrence	Observed probability	Suggested Occurrence probability (logarithmic scale)
static structural elements	0.0093 %	0.0100 %
oil-hydraulic structural systems	0.079 %	0.100 %
re-assembled structural elements	0.368 %	1.000 %
oil-hydraulic mechanical systems	0.632 %	1.000 %
powered electric mechanical systems	0.686 %	1.000 %
human error	7.133 %	10.000 %

- only projects with at least two engineers involved are considered valid for analysis
- for each project, the percentage of proper supervision for each engineer involved is evaluated
- only projects where the sum of the percentages of proper supervision for all engineers is at least 50 % are considered valid for the analysis of the effects of equipment failure
- only projects where the sum of the percentages of proper supervision for all engineers is at least 250 % are considered valid for the analysis of the effects of human error
- to evaluate the number of observed operations for equipment failure effects, all operations in valid projects are considered
- to evaluate the number of observed operations for human error purposes, only 10 % of total operations in valid projects are considered (considering the 'close supervision stages') to evaluate the number of occurrences, this value is defined as an upper threshold number of occurrences in the records of at least three engineers with participation in each project.

Considering these assumptions, the results presented in Fig. 4.1 and 4.2 were observed for each item presented in Table 4.9. Each graph has two scales, one for the probability and the other for the performed operations.

The summary of this study is presented in Table 4.9, where 'safe' probabilities of failure or human error are included in a logarithmic scale. These are coherent with the values presented in Table 4.4.

4.1.3.3 Consequence and Impact Evaluation

In sum, the scale of consequences is defined for a maximum permissible risk of 1 (convention), which is obtained considering the classic probability of structural failure of 10^{-4} times the consequence of a general failure (10^{4}). The other consequence values are obtained precisely in the same way by crossing the typified logarithmic scale probabilities with the consequences that lead to maximum permissible risk.

For simplicity, the criterion presented in this chapter is the measurement of the consequence in terms of injuries and task interruption time, but the method is applicable to any scale of consequences.

Ideally, the criteria should have the prior consent of the operator and the contractor. It should also be noted that the consequence will largely depend on the adopted strategy in terms of preventive measures – downtime will depend, for example, on the available spare parts.Table

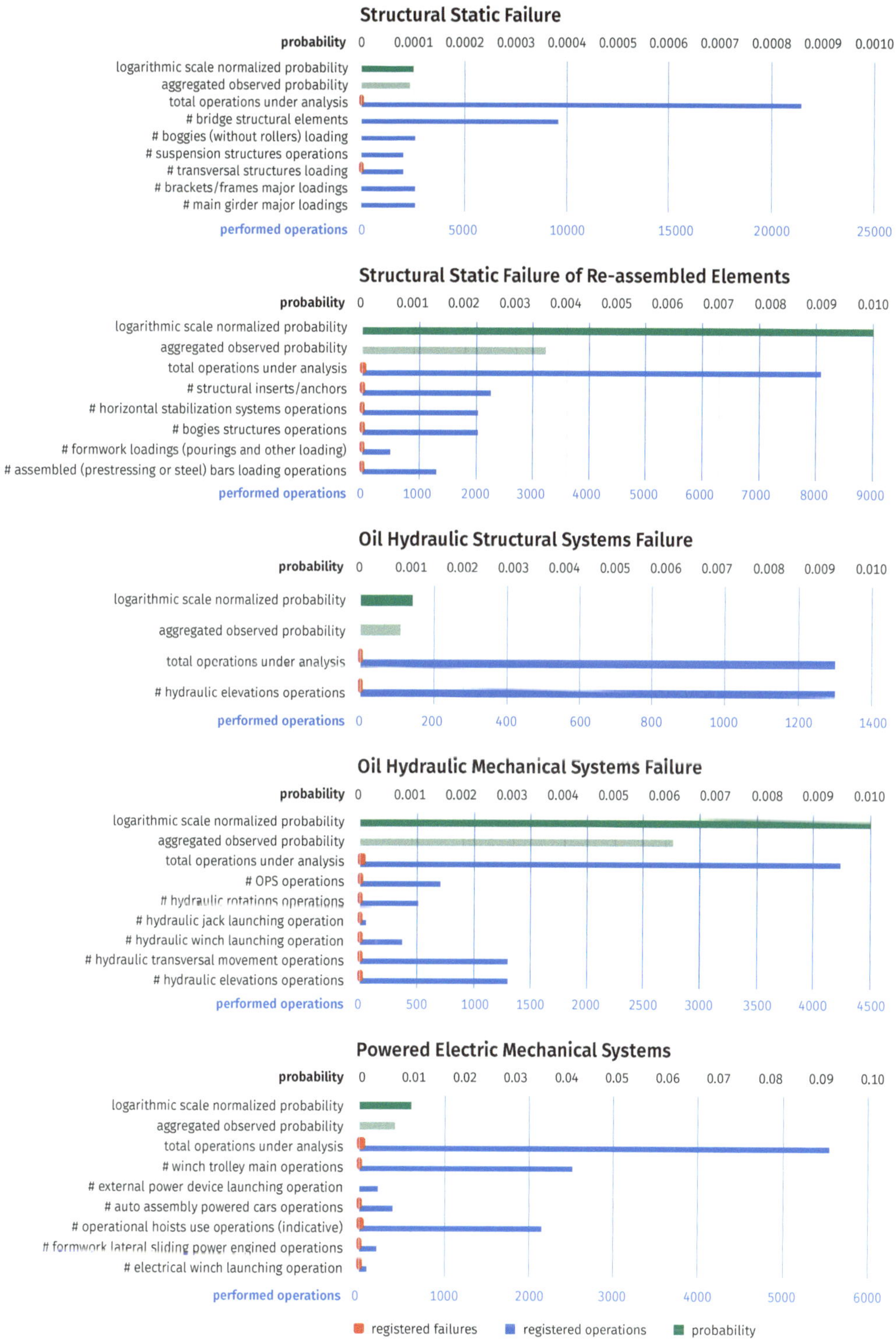

Fig. 4.1 Records and probabilities of different types of equipment

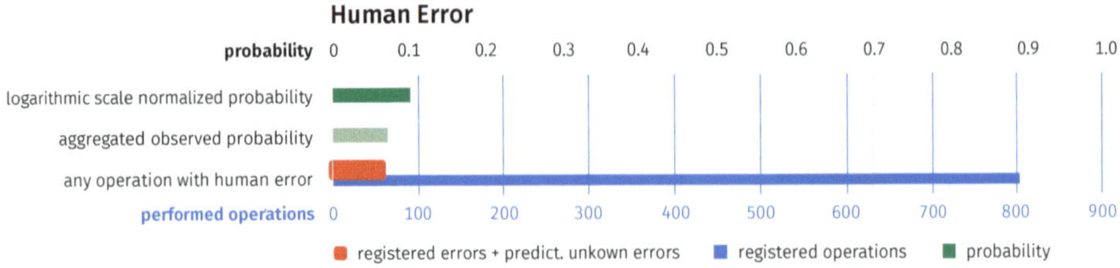

Fig. 4.2 Records and probabilities of human error

4.10 presents a suggestion for establishing consequence levels.

The reference of consequence magnitude can also be evaluated, for example, in terms of costs, on a logarithmic scale. That evaluation would only require appraising the costs of the consequence in one scenario, for example, the cost of a general collapse, and then applying a reverse logarithmic scale for other scenarios. In that evaluation applying the present methodology[1], the acceptable risk would be that reference evaluated cost times 10^{-4}.

4.1.3.4 Risk Evaluation

Considering the definitions presented above, the risk scale indicating the order of magnitude of the risk may be constructed as illustrated in Table 4.11.

4.1.4 Lessons Learnt and Recommendations

A strong recommendation is made to any engineer with responsibility in bridge construction to start using numerical risk analysis, at least as an additional tool to provide reliability. Current non-numerical methods are empirical and subjective and do not ensure systematic and consistent red flags when risk analysis is performed.

The authors have been developing a bridge construction accident database, and it is quite obvious that this type of methodology would have prevented most of the observed accidents. The absence of numerical risk analysis methods in construction is probably the major cause of accidents worldwide.

Using the method presented, once high-risk situations are identified, risk mitigation can be achieved by addressing either the 'probability side' or the 'consequence side' of risks. It is more common to focus on the probability side. Measures often include 'adding simultaneous facts'—conditions that must occur together to induce a consequence, thereby reducing its probability. This can be achieved through the classic rule of redundancy of elements, the use of monitoring and warning devices, physical barriers, locks, or limits to prevent occurrences, or the implementation of additional controlling procedures. Acting on the 'consequence side' is typically more demanding, especially in later project stages, as it may require conceptual changes. It is crucial to be aware of 'red flags' to respond effectively on one or both fronts.

1 (*) To obtain a very indicative intuitive evaluation of consequences in terms of costs, the consequence magnitude order, considering market values of 2023, it can be assumed that the cost of the reference consequence scale (C(O) =1) is 100 (similar in US dollars or euro). Using the scale presented in Table 4.10 a rough estimation of the reference cost of the consequence would be C(O) multiplied by 10^2 (e.g. a C(O)=100 would have a cost of 100 × 10^2 = 10000 currency units).

Table 4.10 Consequence scale

C(0)	Classification criteria: process impact
0	Intervention required without consequences to normal operation.
1	Downtime < 30 minutes.
10	Downtime < 180 minutes.
100	Light injury / Significative malfunction. Downtime < 24 hours.
1000	Severe injury / Significant damage. Downtime < 10 days.
10000	Accident involving loss of human life / Severe incident.

Table 4.11 Risk scale

		Consequence					
		0	1	10	100	1000	10000
Probability	1	0	1	10	100	1000	10000
	0.1	0	0.1	1	10	100	1000
	0.01	0	0.01	0.1	1	10	100
	0.001	0	0.001	0.01	0.1	1	10
	0.0001	0	0.0001	0.001	0.01	0.1	1

<0.01	Very minor risk
0.01	Minor risk
0.1	Moderate risk
1	Maximum acceptable risk
>1	Intolerable risk

References

[1] Soares, J.; Pacheco; P., Coelho, H; Resende, A.; Carvalho, D. and Torres, A., General approach to safety and risk management in bridge construction, IABSE Guimarães 2019.

[2] Soares, J.; Pacheco; P., Coelho, H; Resende, A.; Carvalho, D. and Torres, A., Structural design, and globalization – the particular case of bridge construction equipment, IABSE New York 2019.

[3] European Standard, EN1990:2002. Eurocode 0. Basis of Structural Design. European Committee for Standardization (CEN). 2002.

[4] European Standard, EN 1991-1-4 Actions on structures – Wind actions, Brussels: CEN, 2005.

[5] European Standard, EN 1991-1-6 Actions on structures – Actions during execution, Brussels: CEN, 2005.

[6] IAP-11 – Instrucción sobre las acciones a considerar en el proyecto de puentes de carretera, Ministerio de Fomento de España, Spain, 2012

[7] Manual de Cimbras Autolanzables, 2nd edition, CNC, 2015.

[8] European Standard, EN 1993-1-1 Design of Steel Structures – General rules, Brussels: CEN, 2009.

[9] European Standard, EN 1993-1-5 Design of steel structures – Plated structural elements, Brussels: CEN, 2006.

[10] European Standard, EN 1993-1-8 Design of Steel Structures – Design of Joints, Brussels: CEN, 2009.

[11] European Standard, EN1993-1-10 Design of steel structures – Material toughness and through-thickness properties, Brussels: CEN, 2005.

[12] European Standard, EN 1993-1-11 Design of steel structures – Design of structures with tension components, Brussels: CEN, 2006.

[13] European Standard, EN 1090 – Execution of steel structures and aluminium structures – Part2: Technical requirements for steel structures, Brussels: CEN, 2011.

[14] Resende A., Coelho H., Pacheco P., Preliminary assessment of wind actions in large span MSS, Multi Span Large Bridges Conference, Porto, Portugal, 2015.

[15] Coelho H., Torres A., Pacheco P., Moreira C., Silva R., Soares J. M., Pinto D., Fatigue design and prevention in Movable Scaffolding Systems, Civil and Environmental Engineering Reports (CEER), 2017.

[16] Sedlacek, G., Feldmann, M., Kühn, B., Tschickardt, D., Höhler, S., Müller, C., Hensen, W., Stranghöner, N., Dahl, W., Langenberg, P., Münstermann, S., Brozetti, J., Raoul, J., Pope, R., Bijlaard F.: Commentary and worked examples to EN 1993-1-10 "Material toughness and through thickness properties" and other toughness oriented rules in EN 1993, JRC Scientific and Technical Report 2008.

[17] Pacheco P., Resende A., Torres A., Guerra A., Coelho H., Design of movable scaffolding systems for seismic events – a practical case, IABSE Congress, Christchurch, 2021.

[18] Council Directive 89/392/CE of 14/06/1989.

[19] ISO12100:2010. Safety of machinery – General principles for design – Risk assessment and risk reduction. ISO. 2010.

[20] HSE. Preventing Catastrophic Events in Construction, London, UK, Health and Safety Executive (HSE), Report RR834. 2011.

[21] Thommesen, J., Andersen, H. B., Human Error Probabilities (HEPs) for generic tasks and Performance Shaping Factors (PSFs) selected for railway operations., DTU Management Engineering. DTU Management Engineering Report No. 3.2012.

[22] Kirwan, B. 1994, A guide to practical human reliability assessment Taylor & Francis.

[23] Hickling, N. 2007, An Independent Review of a Rail-specific Human Reliability Assessment Technique for Driving Tasks, RSSB.

[24] Kirwan, B., Gibson, H., Kennedy, R., Edmunds, J., Cooksley, G., & Umbers, I. 2004, "Nuclear action reliability assessment (NARA): A data-based HRA tool", Probabilistic Safety Assessment And Management, VOL 1–6 pp. 1206–1211.

[25] Gibson, H., Hickling, B., & Kirwan, B. 2006, Feasibility study into the collection of human error probability data, EUROCONTROL, Brétigny-sur-Orge Cedex FRANCE, EEC Note No. 02/06.

https://doi.org/10.2749/sp001.ch04.2

4.2 Applicable Standards and Action During Construction

Javier Muñoz-Rojas Fernández
Spain

4.2.1 Relevance of the Construction Design

Design specifications are frequently service design oriented with minimal or limited guidelines with respect to construction. However, the analysis of the construction stage is an important task during the bridge design process. In fact, the differentiation between the construction and design process is, to a certain extent, artificial since both aspects are part of the same problem. The design of any structural solution cannot be addressed without considering how it is going to be erected. This is particularly pertinent in long-span or special bridges where the solution itself or many details are conditioned by the erection sequence.

The relevance of this topic must be carefully considered by bridge designers during the development of their project, and involves dealing with different issues:

- The definition of an erection sequence compatible with the bridge design
- The evaluation of the permanent and transient loads to be considered during the construction stages and the way to combine them. Among the transient loads, there is a special group that will act only during the construction process, being related to the erection itself: equipment, materials storing, temporary elements, induced constraints, etc.
- Finally, the structural verifications of stability, capacity, stresses, reactions, deflections, etc. during the construction stages, which may include particular load combinations.

Each of the topics is discussed in the following sections.

It must be stressed that if the design engineer is not familiar with or has insufficient knowledge of the erection process planned for the project it is highly recommended that the design engineer, in addition to exhaustive structural verifications, consult contractors experienced in bridge erection and/or suppliers of specialised equipment. Accurate information and details of the construction loads should be considered in the structural design to reduce uncertainties and risks in the design. This is not only necessary for the safety of construction but also for the competitiveness and viability of the project.

It should not be forgotten that in the end, the design engineer's primary responsibility is to ensure the proper management of the project's construction risks. The best way to do this is through a profound understanding of the structure and the forces that it must resist at any given time, regardless of the construction or operation conditions. In addition to other bibliographic references, the following documents are key in this matter [1–3].

4.2.2 Erection Sequence

All projects should include a reference to the erection sequence considered in the bridge design to which the structural capacity and stresses of the structure are adapted. For clarity, this should also summarise the alternatives compatible with that design and possible limits of acceptable modifications that will imply changes to the structural design/capacity and the need for new verifications.

Design specifications generally address this topic in a vague way as there is no clear position among consultants and official engineers about the convenience and limits of permitting alternate construction methods (as exemplified in Figgure 4.3) [4–6].

In general, contractual requirements related to construction are established in documents other than design specifications such as in employer requirements or other contract documents. In turn, technical details are also normally addressed during the construction stage with the involvement of the contractor and are incorporated in documents such as erection manuals, erection procedures, or similar [13].

Specifications are generally vague about the designer's work. A good example of this approach is revealed in the provisions of the AASHTO LRFD Bridge Design Specification Art. 2.5.3:

> *"Bridges should be designed in a manner such that fabrication and erection can be performed without undue difficulty or distress and that locked-in construction force effects are within tolerable limits. When the designer has assumed a particular sequence of construction in order to induce certain stresses under dead load, that sequence shall be defined in the contract documents. Where the bridge is of unusual complexity, such that it would be unreasonable to expect an experienced contractor to predict and estimate a suitable method of construction while bidding the project, at least one feasible construction method be indicated in the contract documents."*

This latitude is reasonable in many cases – otherwise, it could restrict free competence or exclude the contractors from taking advantage of their experience and expertise – but it cannot be ignored that from the bridge designer's and even the client's perspective, it may bring uncertainty as to the responsibilities/liabilities resulting from the final construction, particularly in engineering, procurement, and construction projects if the initial designer is not involved in the construction stage and the erection method is modified. For these reasons, it is the responsibility of the design engineer to clearly state in the contract documents, especially in the drawings and the structural analysis report, the assumptions considered in the design and the assumed erection method with the limits where modifications will impact the capacity or integrity of the structure.

It is important to be as clear as possible when establishing the limits of the validity of a design and to highlight when deviation from the planned construction method may impact the capacity to resist loads at least equal to those used in the design.

After a value engineering redesign, the engineer responsible for the new study should assume full responsibility for the project (Engineer of Record). In this

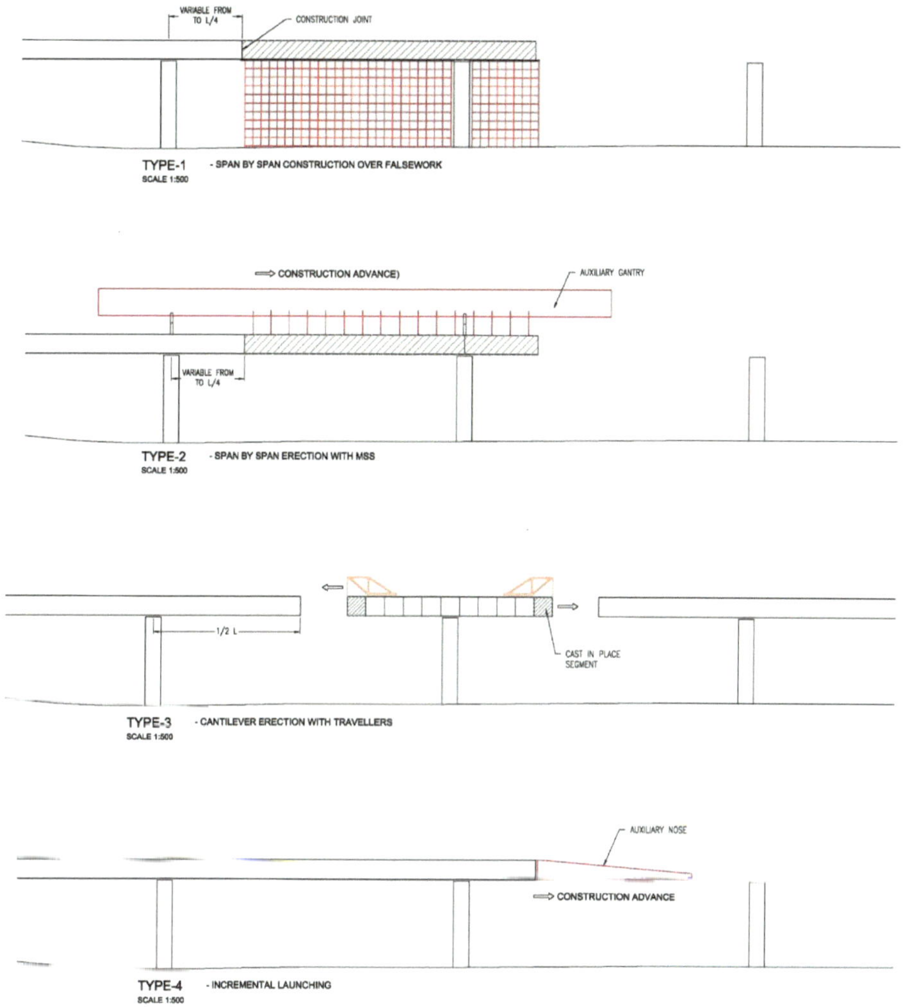

Fig. 4.3 Schematic representation of different in-situ erection methods of concrete bridges resulting in a similar structural configuration

regard, AASHTO LRFD Bridge Design Specification Article 5.12.5.5 is one of the few references that addresses this dilemma openly. Though this is in the context of concrete segmental design, its provisions and comments are extendable to any construction method [8].

From a technical point of view, erection sequences should be analysed carefully in the design stage, particularly in projects with evolutive construction (free cantilever, incremental launching, span-by-span, etc.). Changes in the structural or sectional configuration should be considered carefully with subsequent verification of the capacity, stability, and stress levels of the different elements. Also, as noted above, the bridge designer should clearly refer to the construction conditions compatible with its design.

4.2.3 Structural Verifications and Design Criteria

4.2.3.1 Design Philosophy

All modern bridge design specifications follow a probability-based philosophy in which the uncertainty when establishing the representative values of the actions to be considered in the design is addressed assessing larger or smaller values for the actions according to their probability of occurrence. In turn, the probability of a simultaneous presence of different loads is considered when combining these loads by means of different load factors.

The analysis of the structure during construction is no exception to this and follows the same approach. The particular conditions experienced by the structure during the construction stage and the low return period that may be assumed in some transient actions are logically reflected in the load factors and in the way to combine loads that may differ from the final analysis of conditions.

Bridge construction should be checked to ensure that structural capacity is assured throughout the entire construction process, that damage does not occur, and that construction methods and equipment have no adverse impact on the service condition or life of the completed structure.

In turn, each construction process introduces some specific problems and may require verifications – such as friction forces during the launching process, accidental situations in free cantilever erection, etc. – the criteria or limits of which may differ from the final structural system of the bridge. Some general features are presented in the subsequent sections.

4.2.3.2 Stress Levels

Design specification generally accepts higher limits for temporary stresses during construction. AASHTO Load and Resistance Factor Design (LRFD), particularly Table 5.14.2.3.3-1 Review of Relevant International Standards (AASHTO, EN, PTI) points to the following indicative values:

Concrete

- Compression stresses. The main design objective is to prevent levels that may induce inelastic deformations. In service limit state (SLS) combinations, compression stress limits are increased to values between 60 % and 65 % of the compression strength
- Tensile stresses. Provided there is enough reinforcement for the tensile force computed assuming an uncracked section, stresses between 50 % and 60 % of the tensile strength are admissible. [Table 4.1]

Cables

- According to PTI and in SLS combinations, maximum stresses in stay cables may reach 60 % guaranteed ultimate tensile strength (GUTS) in systems tested for axial fatigue with bending effect, or 55 % GUTS in non-tested systems. Major public-private institutions [13] and European standards provide similar values.

4.2.3.3 Concrete Cracking During Construction

In concrete structures crack effects during construction should be analysed with the same criteria used for service conditions. However, the possible existence of tem-

Table 4.1 14.2.3.3-1 AASHTO LRFD. Tensile Stress Limits in concrete during construction

Load Combination	Stress Limits			
	Flexural Tension		Principal Tension	
	Excluding "Other Loads"	Including "Other Loads"	Excluding "Other Loads"	Including "Other Loads"
a	0.190√fc'	0.220√fc'	0.110√fc'	0.126√fc'
b	0.190√fc'	0.220√fc'	0.110√fc'	0.126√fc'
c	0.190√fc'	0.220√fc'	0.110√fc'	0.126√fc'
d[1]	0.190√fc'	0.220√fc'	0.110√fc'	0.126√fc'
e[2]	0.190√fc'	0.220√fc'	0.110√fc'	0.126√fc'
f[3]	0.190√fc'	0.220√fc'	0.110√fc'	0.126√fc'

porary cracks may be acceptable if subsequently closed by additional prestressing or normal forces, as this situation has no major impact on the durability and appearance of the concrete.

Cracking pattern and width should be anticipated through suitable structural analysis. Obviously, uncontrolled cracking must be prevented and any unintended cracks emerging during construction must be investigated and rectified after construction.

4.2.3.4 Geometrical Control and Precambers

The evaluation of construction deflections is another important topic to be addressed during the design process, especially for – but not limited to – long-span bridge projects and/or solutions with an evolutive erection method. Long-term effects must be logically considered in concrete and composite structures. The anticipated deflections, rotations, precambers, or erection positions should be reflected in the contractual documents, as well as the construction tolerances and permitted deviations. It is also crucial to include mitigation strategies for out-of-tolerance components.

Frequently, the final erection sequence is modified by the contractor and/or adapted to its construction equipment. In this case, the figures of the reference design must be reevaluated by the contractor's technical officers or the supporting engineers.

4.2.4 Actions

4.2.4.1 Generalities

In the design stage, the distribution and application of loads during the construction phase should be selected to produce the least adverse effects.

4.2.4.2 Permanent Loads (Dead Loads)

As it is known, permanent loads do not change in magnitude and are in general caused by gravity loads. They should be determined according to the provisions of applicable specifications. In general, they are spatially fixed and do not introduce dynamic amplifications – apart from the inertial effects due to seismic or wind actions, that are considered in the dynamic analysis of these actions.

During the structural analysis of the construction, the permanent load should be introduced in the structural models according to the planned erection sequences. Combinations and load factors will follow the values stated in the applicable specification. In some verifications, load factors may be smaller than those used in the final structural conditions.

Realistic self-weights should be used to ensure the final deck geometry will be correct. An exaggerated concrete density or section dimensions are safe and conservative for capacity and stability verifications but may lead to an unrealistic geometry and unconservative locked-in permanent load effects and creep redistribution.

4.2.4.3 Transient Loads

Due to the limited duration of the construction compared to the design life of the structure, the probability return periods to establish the representative values of some transient actions may be reduced during erection leading to reduced characteristic values.

The most critical transient loads for construction verifications are generally due to wind, seismic, and thermal action. Logically other transient loads, like water loads, vessel impact, etc., may govern the design in particular conditions. In those cases, and similarly to other transient loads during construction, a reduced probability may be assumed to establish the values or load factors of these actions to be used in the verifications.

Wind loads

The characteristic values of wind velocity during construction are established in some cases by the client in the Employer Requirements. Otherwise, the provisions of the applicable standard design specifications and/or some specific guidelines, such as AASHTO, should be used. Guide specifications for wind loads on bridges during construction provide values for the wind during construction, or a method should be used to evaluate it from the maximal wind velocity considered in the design of the completed structure [7].

It is generally accepted that a lower velocity due to the reduced return period of the action may be applied according to the construction time anticipated in the design. As a reference value for construction periods between one and three years, the resulting return period ranges between four and ten years.

The Eurocode approach establishes in EN-1991, Section 1-4, a methodology to evaluate a reduction factor taking into account the actual construction period. The 10-minute wind velocity with probability p for an annual exceedance can be determined by multiplying the basic wind velocity (50-year return period) by the following factor:

$$c_{prob} = \left(\frac{(1 - k \cdot ln\, ln\, (-\, ln\, ln\, (1 - p)))}{(1 - k \cdot ln\, ln\, (-\, ln\, ln\, (1 - 0.02)))} \right)^{n}$$

Assuming a 3-year construction period with a 10-year return period (p=0.1) and considering k=0.2 and n=0.5, results in a reduction factor of c_{prob}=0.902.

Another aspect to be considered in wind analysis during construction is that sections and surfaces exposed to wind may be different from the conditions after construction completion, therefore modifying the pressure coefficients from those considered in the general design or those obtained in wind tunnel tests.

Wind should also be applied to the construction equipment, the presence of which adds more exposed surfaces, possibly also modifying the wind flow and hence pressure coefficients.

Dynamic actions due to wind excitation during construction should also be analysed if the structure is susceptible to vibrations. It should also be considered that sectional and dynamic properties may differ significantly from the final conditions and that construction loads and/or equipment may also have an impact.

Wind during erection may also be critical for the construction equipment design or for its working conditions. Displacement of special gear (travellers, movable scaffolding systems, gantries), special manoeuvres (arch tilting, heavy segment lifting, heavy lifting operations), operation of cranes, etc. are normally restricted to low wind conditions, the velocity of which should be specified in the operations manual. Operations should thus be scheduled according to wind velocity forecasts of meteorological services. In excessive wind conditions, the equipment must be secured by suitable arrangements which may need to be anchored to the structure (Fig. 4.4) [12].

Seismic loads

Similarly to wind, the seismic action during construction may be established by the client or based on an analysis considering a reduced return period of the event. In some cases, particularly in design-build projects, the contractor may also decide the willingly incurred risk level.

Fig. 4.4 Typhoon Odette (NOAA/CIRA/RAMMB, CC-BY 3.0) over the Philippines during construction of the Cebu-Mactane cantilever segmental bridge (250 km/h)

Given the, in general, intrinsic low probability of maximum earthquake movements, design specifications admit the possibility of reducing seismic actions during construction.

AASHTO Guide Specifications for LRFD Bridge Seismic Design [2] indicates in Section 3.6 (Temporary and Staged Construction) that bridges constructed in stages in periods smaller than 5.0 years may be designed with a reduced elastic response spectrum. A maximum reduction factor of 2.5 for the extreme event action is suggested.

This provision should be considered carefully by authorities and designers, and evaluated considering the risks of each project. If the collapse of the erected structure results in a catastrophic scenario – which may be the case for bridges over heavy traffic ways – or there is a significant probability of loss of life in that event, more conservative assumptions should be considered. In general, a reduction factor of 2.0 is a normal assumption in projects based on AASHTO Specifications.

Eurocodes follow the same philosophy of reducing the seismic event due to its lower probability of occurrence during construction but based on a more detailed methodology. Annex A of EN-1998-2 includes detailed guidance for the selection of the seismic action during construction [11].

It is based on assuming that the return period of the seismic event (T_{Rc}) depends on the time of the construction phase (t_c) divided by the probability of exceedance of the design seismic event (p) which should not exceed 0.05. The reduction factor of the reference peak ground acceleration – corresponding to the extreme event with the return period T_{NCR} equal to 475 years – is then evaluated by $(\frac{T_{Rc}}{T_{NCR}})^k$, with k in the range 0.30–0.40.

A practical example with $t_c = 5$ years, $p = 0.05$, brings $TRc = 60$ years and a reduction factor $(\frac{100}{475})^{0.4} = 0.30$, close to the 0.40 factor obtained with the AASHTO approach.

Snow loads

In regions of heavy snowfall, this load may be critical during construction. The values to be considered depend logically on the region's pluviometry. In areas of the North United States of America and Canada, loads of 0.5 kPa – equivalent to about 1 metre of depth of snow – are frequent. Here, provisions are usually set establishing the need to remove the snow if it exceeds a given depth (normally 1.2 metres).

Thermal loads

Thermal loads during construction should be considered assuming the same values established by the applicable standard for service conditions. In evolutive structures, the effects of temperature are particularly important for geometrical control during construction.

In some cases, it is also important to establish temperature conditions for operations in which the restriction to expansion or contraction may induce additional forces. Installation of bearings and joints may require the same precautions in the design documents.

Long-term effects (creep and shrinkage of concrete)

Analysis is required of the effects of creep and shrinkage in concrete and composite structures to establish the geometrical conditions during construction and operation, and for assessment of redistribution of reactions and stresses that they induce.

Friction loads

Connected to the two previous loads, the effects of resistance to temperature and imposed deformations should be considered during construction based on representative values.

4.2.4.4 Construction Live Loads

Generalities

These loads cover the weight of construction materials or auxiliary equipment

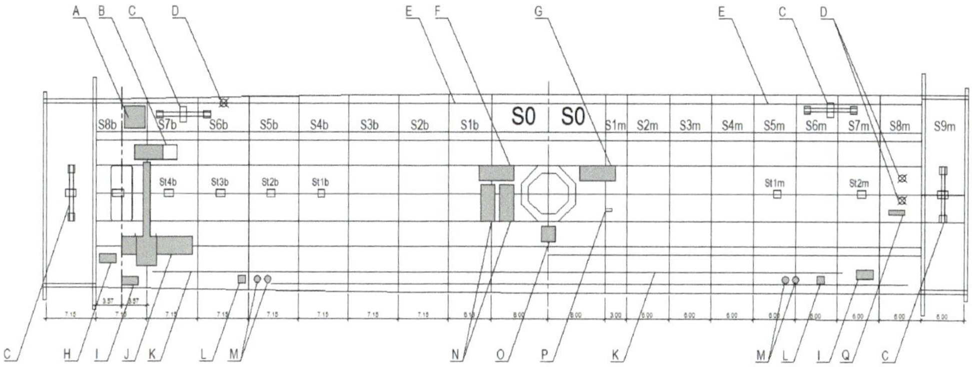

A Rebar base and walls
B Truck Crane, 26tn
C Delta frame with anchorage
D Mechanical distributor, 1tn
E Concrete tube, 25kg/ml
F Power generator 400kv, 4tn
G Water tank, 4tn
H PT stressing device, 2tn
I Tools for rebar and delta frame, 4tn
J Mobile crane 50t, 32tn
K Stay cable preparation
L Pushing device, 1t
M Coil, 2 × 3.5tn
N Containers, 4tn
O Access hoist – Alimax, 10.5t
P Grouting machine, 1tn
Q Compressor, 4tn

Fig. 4.5 Map of construction loads in a cantilever construction of a cable-stayed bridge.

stored on sections of the previously built structure, i.e., reinforcement bars and post-tensioning, special equipment or devices for erection, cranes, jacking pumps, etc (Fig. 4.5 and 4.6.

The values to be adopted depend on the construction method and no general rules can be prescribed. The bridge designer's experience or consultation with specialists may provide the required inputs for the design stage. Final precise values are normally stipulated during the construction stage in erection manuals or equipment supplier specifications [9, 10].

AASHTO LRFD (9th ed.), Section 5.12.5.3.2, includes a detailed description of different construction loads. Although directed at concrete-segmental bridges, the provisions and methodology offer a useful tool to address other construction methods. According to its notation, a distinction is made between distributed construction live loads like plant, machinery, materials- and special construction equipment-traveller, or gantries.

It is recommended that the erection manual or similar documents prepared for the construction stage include a construction loading table that should be respected, particularly for the geometrical control survey.

Construction live loads

In general, these loads are often not accurately known – in terms of both magnitude and location – at the time of the design, but in many cases, it is necessary to establish preliminary values for structural or stability verifications.

If specific loads have been assumed as a part of the design, then these loads should be documented in the plans or project documents.

Otherwise, during construction, it will be necessary to check or evaluate the structure's actual capacity so to determine the magnitude and permitted positions. This input should then be documented in plans as well as, if necessary, the definition of the

Fig. 4.6 Cantilever construction of a cable-stayed bridge.

temporary supports or auxiliary elements to ensure stability and constructability.

In addition to the load value, precise control of live loads is necessary for satisfactory geometrical control during bridge construction. A plan with the precise position of each load will be established for each construction stage, so there is no need to add this distributed construction live load.

Specialised equipment loads

These elements include specialised gear used to temporarily support the structure or to erect/lift its main elements (traveller, gantry, lifting frames, or similar) and other auxiliary elements (cranes, delivery dollies, etc.) but may also involve relatively heavy elements that are frequently supported or hoisted on the previously built structure (i.e., travellers, launching gantries, lifting frames, etc.).

These elements can turn critical in the design – like a free cantilever with travellers (Fig. 4.7). The concentrated forces they transmit to the structure have to be accurately defined (magnitude and position) so that the corresponding verifications (overall or locally) can be carried out.

4.2.5 Particular Design Requirements for Special Construction Methods

4.2.5.1 Cantilever Construction and Incremental Launching

Cantilever construction is one of the more versatile bridge construction methods, efficiently and economically applicable to spans ranging from 100 metres (concrete segmental bridges) to 1000 metres (steel or composite cable-stayed bridges).

It is also a good example where the erection method strongly conditions the structural configuration and elements of the bridge, and where the most demanding situations may happen during the construction stage.

For that reason, some design specifications or design guidelines address this construction method in detail.

AASHTO LRFD Bridge Design Specifications [1], Section 5.12.5., includes thorough provisions for segmental concrete bridges. Subsections 5.12.5.3.2 through 5.12.5.3.5 refer to construction loads and their combinations and verifications (Table 4.2).

The SETRA design guide on prestressed concrete bridges built using the cantilever method [13], presenting an overview of the French experience in these kinds of bridges, also includes a section on the cantilever stability during construction and the loads and situations to consider.

These references provide information on several load combinations during construction stages for verification of the equilibrium of the cantilevers. Various unbalanced situations are considered assuming differential values of the self-weight of each cantilever, staggering the segment construction, different construction loads,

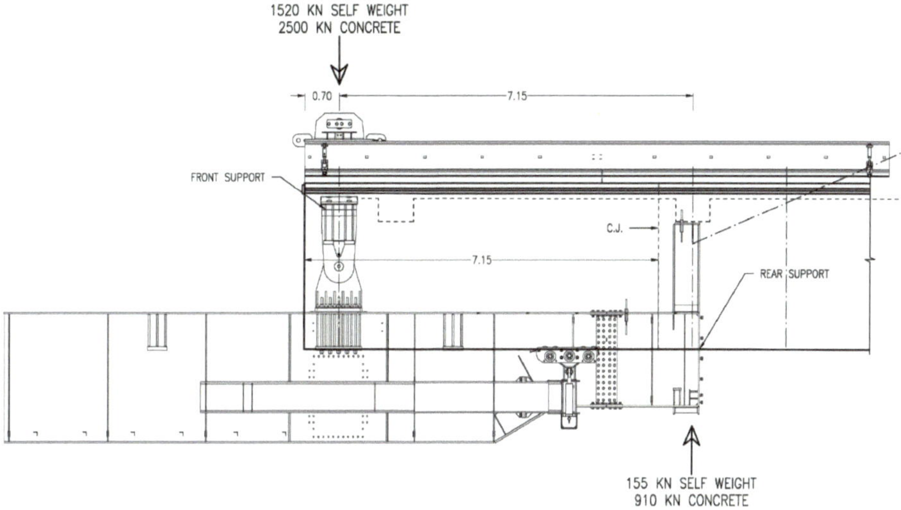

Fig. 4.7 Scheme of reactions transmitted by an underslung traveller in cantilever construction

Table 4.2 AASHTO LRFD. Segmental bridges. Load Factors and Load Combinations during construction

Load Combi-nation	Load Factors														
	Dead Load			Live Load				Wind Load			Other Loads				
	DC	DIFF	U	CEQ	CLL	IE	CLE	WS	WUP	WE	CR	SH	TU	TG	WA
a	1.0	1.0	0.0	1.0	1.0	1.0	0.0	0.0	0.0	0.0	1.0	1.0	1.0	γ_{TG}	1.0
b	1.0	0.0	1.0	1.0	1.0	1.0	0.0	0.0	0.0	0.0	1.0	1.0	1.0	γ_{TG}	1.0
c	1.0	1.0	0.0	0.0	0.0	0.0	0.0	0.7	0.7	0.0	1.0	1.0	1.0	γ_{TG}	1.0
d[1]	1.0	1.0	0.0	1.0	1.0	0.0	0.0	1.0	1.0	0.7	1.0	1.0	1.0	γ_{TG}	1.0
e[2]	1.0	0.0	1.0	1.0	1.0	1.0	0.0	0.0	0.0	0.3	1.0	1.0	1.0	γ_{TG}	1.0
f[3]	1.0	0.0	0.0	1.0	1.0	1.0	1.0	0.0	0.0	0.3	1.0	1.0	1.0	γ_{TG}	1.0

vertical wind, and accidental cases in the event of the collapse of a form traveller or fall of a precast segment (Fig. 4.8).

Incremental launching is another good example where the erection method conditions the structural configuration and details of the bridge, and, hence, where the construction design is more tightly connected with the general design of the bridge. The system is used both for concrete and steel solutions. From the structural standpoint, these structures have the particularity that most sections undergo the maximum positive and negative bending moments during the launching process.

The structural design of the bridge requires the analysis of all the construction stages (Fig. 4.9) and local effects in regions with concentrated forces during the launching manoeuvres.

Some elements and equipment for the construction require specific analysis (e.g., casting yard normally connected to one of the abutments, auxiliary steel nose in concrete bridges to reduce the free cantilever length of concrete, etc.).

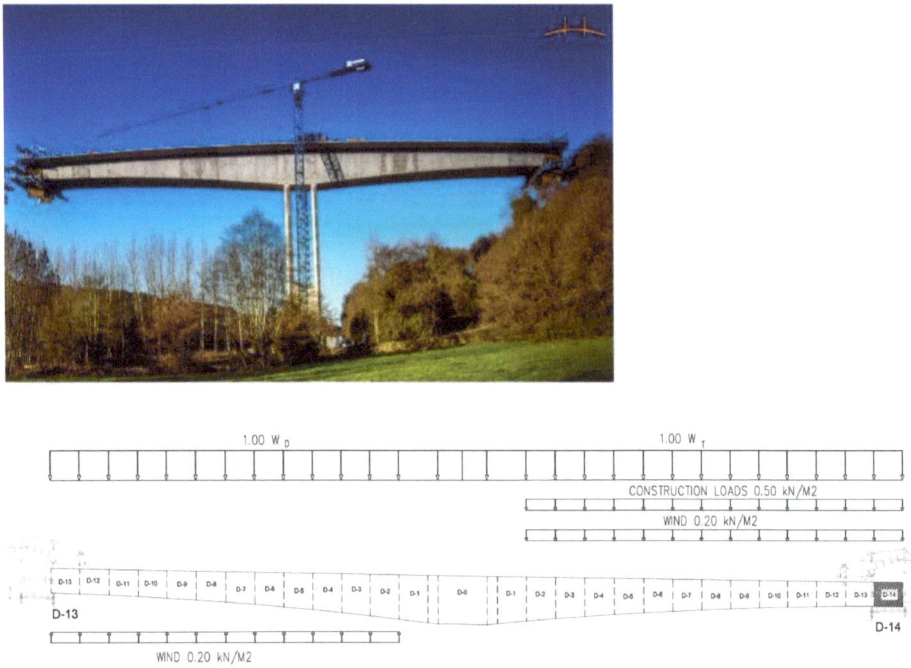

Fig. 4.8 Typical unbalanced situations considered in cantilever construction 1. Differential self-weight + wind + segment erection. 2. Accidental failure of traveller.

For the superstructure design, frictional forces on bearings should be evaluated carefully. Friction coefficients of Teflon or similar materials are around 3%. In static conditions before sliding, this may be significantly higher and should be considered in the design. The positive effect of the friction forces should not be considered. Auxiliary anchoring may be necessary in tall piers to control horizontal forces in pier heads. Provisions should also be made to restrain the girder if it is launched with a gradient.

4.2.6 Lessons Learnt and Recommendations

- In design, it is good practice to interact with bridge construction equipment suppliers to evaluate construction stage actions on the bridge structure. Following the definition of the load maps, a discussion on accidental loads and/or dynamic loads (as breaking forces, etc.) is highly recommended.

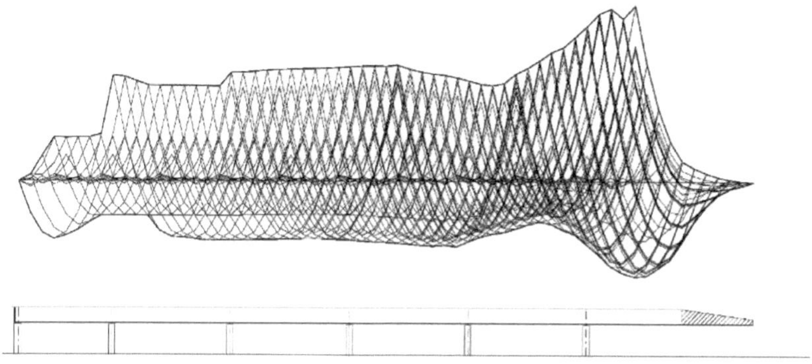

Fig. 4.9 Typical bending moments in a concrete girder during an incremental launching construction

- The development of bridge equipment interaction models is fundamental to providing safe structural analysis and bridge design. Both in span-by-span precast segmental construction and in span-by-span in-situ construction, the development of bridge equipment interaction models should be mandatory and this responsibility should be stated in the contract between the constructor and the equipment supplier.

References

[1] AASHTO LRFD Bridge Design Specifications 9th Edition. 2020

[2] AASHTO Guide Specifications for LRFD Bridge Seismic Design. 2011

[3] AASHTO. Guide Specifications for Wind Loads on Bridges During Construction. 1st Edition 2017

[4] AASHTO. Construction Handbook for Bridge Temporary Works, 2nd ed. 2017

[5] European Standard EN 1997-1.2004

[6] European Standard EN 1990:2002. Basis of Structural Design

[7] European Standard EN 1991-1-4:2005. Eurocode 1: Actions on structures. Part 1–4: General actions – Wind actions

[8] European Standard EN 1991-1-6:2005. Eurocode 1: Actions on structures. Part 1–7: General actions – Actions during construction.

[9] European Standard EN 1991-1-7:2006. Eurocode 1: Actions on structures. Part 1–7: General actions – Accidental actions.

[10] European Standard EN 1998-2:2005. Eurocode 8. Design of structures for earthquake resistance. Part 2: Bridges.

[11] European Standard EN 1998-2:2005. Eurocode 8. Design of structures for earthquake resistance. Part 2: Bridges. Annex A

[12] PTI. Recommendations for Stay Cable Design. 6th Edition. 2012

[13] SETRA Design Guide. Prestressed concrete bridges built using the cantilever method 2007

4.3 Cost

Carlos Bajo
Spain

4.3.1 Introduction

Historically, bridge design skills have been taught at technical universities considering just the proper use of materials and structural shapes. Traditionally, cost-effective solutions were considered to be those equivalent to the optimum consumption of materials. However, currently, a new cost-driver with a significant impact on final costs should be considered: this new agent is construction methodology; specifically involving equipment, labour, time, and costs associated with financing and insurance.

In fact, bridge construction experience shows that the adequate decision-making process for construction methods and materials involves a holistic approach to cost evaluation—not just a simple "materials cost evaluation".

The main goal of this chapter is to provide the reader with guidelines and reference topics to develop such a holistic approach. At the end of the day, and as well known, construction methods are deeply influenced by costs, i.e., economic aspects, as well as other fundamental factors (as presented in Section 1.3.1).

4.3.2 Cost Items

As mentioned above, the final cost of an infrastructure project is not only a matter of material cost. For instance, the cost of a cubic meter of concrete will be different if it is poured at ground level versus poured at height under very demanding temporary safety conditions. Moreover, the cost would be different if specialised construction equipment and specialised labour were required to pour this concrete. In this instance, the selection of the construction methodology will undoubtedly affect the project's final cost.

The different cost components that can be considered along the design and built process concern:

- Structural conceptual design.
- Selection of construction methodology.
- Redesign during the detailed design stage for construction within the programme.
- Finishing works.

All these cost items, or stages, require their own time to accomplish the different steps and cannot be overlapped easily among the different stages since each stage is carried out by different agents without direct coordination or a clearly linked relationship between them. Designers, reviewers, builders, and checkers do not always work together harmoniously; the working relationship is often riddled with conflicts and disagreements that cost money and time. The clashes encountered, found mainly during the construction processes, often

Fig. 4.10 Several bridges under construction

involve undefined or missing items in the design that need to be addressed before construction. For example, when the design does not consider the proper weight of a crane on top of the bridge or any other structure, this could involve redesigning the bridge and in some cases a change in dimensions of crucial elements and thus the final cost of the structure.

4.3.3 Known Structural Typologies

Certain bridge typologies have been developed by considering construction means and methods. Some of the methods herein are cost-effective due to the clever use of stability and static equilibrium during construction, sometimes taking advantage of the bridge itself. Balanced cantilever solutions for cable-stayed bridges or suspended bridges, for example, are cases where the typology is intrinsically linked to the construction method effectively using stability and gravity to reduce costs.

Several similar methods have been employed in bridge construction:

- Balanced cantilever construction:
  - Cast-in-place, box girder construction
  - Precast segmental construction
  - Extradosed bridges
  - Cable-stayed bridges.
- Span-by-span construction:
  - Movable scaffolding system (MSS) construction
  - Incremental launching
  - Precast segmental.
- Precast construction:
  - I-girders
  - Heavy precast construction.
- Heavy lifting/transportation

Deployment of these methods, in conjunction with a comprehensive set of controlled staging, is often currently referred to as accelerated bridge construction (ABC).

4.3.4 Common Construction Procedure

In advanced economies such as those of the Organisation for Economic Co-Operation and Development (OECD) countries, even the procurement of small infrastructure projects is a lengthy process that involves numerous stakeholders and diverse contract types.

A typical procurement process is as follows:

- Identification of the need for a new infrastructure
- Political decision of investment, capital allocation, and establishment of timeline
- Hiring of a consultant to carry out preliminary analyses, produce the conceptual design, and draft the preliminary budget
- Initiation of the tender process, selection of contracting type (i.e., hard bid, design and build), and funding method (public-private partnership and other variants).
- Selection of preferred bidder for delivery phase
- Detailed design process, including approval processes
- Construction phase.

Throughout this entire process, and especially at the design stage, some key decisions must be taken without knowing whether they are going to be considered in the final detailed design. Usually, the team involved at each stage tries to optimise their own expenses without fully considering the expenses that their decisions trigger in the following or even previous stages. Consequently, some designs may not consider required temporary works as they are perceived as an external part of construction.

This is a kind of inefficiency inherent to this process because rework (the unnecessary reiteration of design tasks) leads to a cost increase for the whole project and causes project costs to balloon. This can sometimes lead to project cancellation due to a lack of funds or delays.

It is worth noting that for many bridges today, construction time, mobilisation time, and costs associated with means and methods may constitute a greater portion of the total project cost than direct material costs.

In this regard, it can be said that every bridge is essentially a unique prototype. Thus, it is quite difficult to take advantage of industrialised construction unless there is a clear repetitive action suitable for prefabrication or mass production.

4.3.5 Accelerated Bridge Off-Site Construction

Historically, bridges have been built on-site. Nowadays, with the development of heavy-duty equipment capable of transporting a whole bridge, working off-site is a possibility.

Not all the bridge typologies are suitable to be built off-site. The principal issue is that some bridges are not structurally stable for temporary conditions. This is not the case with some typologies such as through arches which can be split into different stabilised parts. The central piece can commonly be treated as a simply supported bowstring, which is a self-stable structure and different pieces can be built outside the final location (Fig. 4.11).

The use of heavy-duty machinery, mainly self-propelled modular transporters (SPMT) or other heavy lifting means, can facilitate the construction offsite (Fig. 4.12).

4.3.6 Cost-Effective Design of Bridges

Overall cost is a key aspect all engineers should keep in mind while designing and constructing bridges and structures.

Fig. 4.11 Heavy lifting of central part of a through arch

Fig. 4.12 Heavy transportation of a central bowstring of a through arch

During the process of raising funds to build expensive infrastructure projects, some items that may be presumed to be small during the decision-making process could lead to a repetition of work and subsequently to dramatic cost escalations.

In terms of bridges, the overall design of a structure should consider what is commonly considered temporary works as an integral part of the main design. Mainly because there is great interaction between the design and layout of the main bridge and temporary means and methods used in construction. An example of this is when the structure itself is used to support the weight of the temporary scaffolding instead of using a proper foundation at ground level, for instance, movable scaffolding systems which are rear-mounted on an already hardened bridge deck.

When it comes to bridge design cost, what was considered optimum from the point of view of material cost may not be the most efficient solution when considering construction means, material cost, maintenance cost, etc. The different costs can be grouped into three main categories:

- **Bridge Design Cost**

  Bridge design cost involves all the engineering costs to develop a project, like geotechnical investigation, consulting fees, etc. Depending on the designer's expertise and background, this stage may consider previous experiences, even construction costs, but that is not common practice.

- **Bridge Construction Cost**

  Construction costs are those that are spent just to place the materials in the configuration specified in design drawings. This cost includes direct material costs, construction means, cranes, travellers, different sorts of formwork, all temporary works required for carrying out the construction, and labour needed to place the materials onsite.

- **Maintenance Cost**

  Maintenance cost refers to any cost incurred to keep the bridge in good working condition or prevent future deterioration. This cost can be considered in advance within the construction cost as capital expenditure (CAPEX) or deferred and considered only during service life as operating expenses (OPEX).

Depending on the type of contract, maintenance costs may be included in the overall bridge cost, depending on the different stakeholders involved:

- **Public-Private Partnership (PPP) Projects**

  Maintenance costs must be included in the bridge cost.

- **Design And Build Projects**

  Bridge design and construction costs are considered together to achieve the most economical design to fulfil a target budget.

- **RED FIDIC Projects**

  Construction is constrained by the previous assumptions provided by the client. No further changes are allowed, as established by the International Federation of Consulting Engineers (FIDIC) in its red book on building and engineering works designed by the employer.

Not all elements impact the final bridge cost in the same way, as many are fixed elements where only limited design freedom can be exercised. Abutments, foundations, piers, and decks are examples of structural elements with limited design freedom, as their design is almost entirely dictated by local site conditions. Aside from bridge superstructures, the rest of the elements imply similar building conditions. Abutments are built with their foundations at ground level and foundations and pier elevation will need ground access at different locations. On the other hand, the superstructure is a unique element that can be considered a differential element in terms of cost. Depending on the span length to be bridged, different solutions can be used granting the designer greater freedom in superstructure design. In bridges, the selected superstructure typology, piers, and foundation types will be the main items influencing the final cost of the bridge.

A chart of the different cost portions involved in bridge construction is presented below, irrespective of the bridge elements:

BRIDGE/STRUCTURE COST ESTIMATION

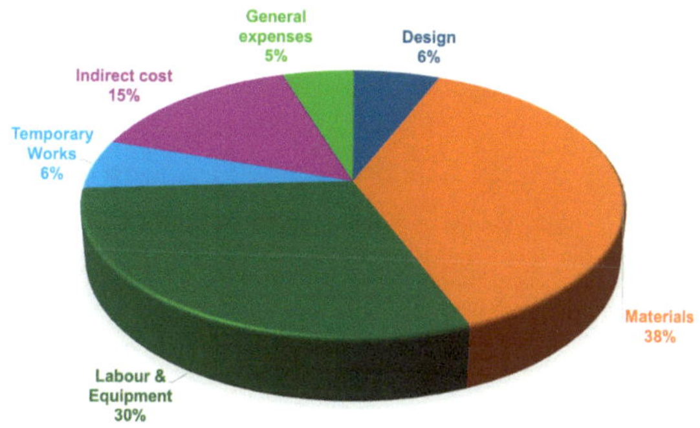

The construction cost breakdown shown above was drawn according to the following criteria:

1. Design: cost of engineering, surveyance, and drafting work associated with the conceptual, tender, and detailed designs, including any site information reports (geotechnical campaign, environmental aspects, utility's location, etc.) and construction site support
2. Construction:
   a. Materials (permanent materials)
   b. Labour and equipment: all equipment, operators and crews needed for construction
   c. Temporary works (access, platforms, scaffolding, sheet piles, traffic management, environmental supervision, etc.)
3. Indirect cost: management staff, compounds, and welfare

4. General expenses: health and safety, quality control, insurance, completion bonds, and contingencies, excluding overheads and profit.

4.3.7 Lessons Learnt and Recommendations

Bridge design will necessarily affect the cost of the final product. In this regard, establishing the different stakeholders and their specific associated costs is crucial to prioritise decision-making and to understand the hierarchy of costs involved.

The typical different stages detected for developing a new project are:

- Establishment of the need for a new infrastructure project
- Securing funds to perform preliminary and feasibility studies
- Conduct of site investigations
- Creation of a preliminary design to establish an initial budget
- Specification of the project's service life
- Determination of the type of contract and funding approach (public-private partnership, red FIDIC project where the employer carries out the design, yellow FIDIC project where the contractor carries out the design, etc.).

Once the project has been approved, the usual different contracting stages are:

- **Tender process:**
  - Additional site investigations
  - Utilities protection or relocation, traffic management, environmental works.
  - Development of a more detailed design (tender design), considering the client's specifications and construction methodology within the contractual framework (completion date, etc.)
  - Identification of temporary works and required design
  - Establishment of the design and construction price.

The type of tender will necessarily affect the method of financing the project and the constraints imposed by the client, authorities, or other entities.

- **Constructor selection:**
  - The process usually starts with a preselection of a reduced number of contractors (shortlist) entitled to submit their proposal. The final contractor is evaluated and selected from these proposals.
- **Construction stage:**
  - Once the client has selected the preferred bidder, the project is awarded. After a conditional award, the client and the preferred bidder enter a negotiation phase during which the terms of financial close are laid out. These terms include the timing of the release of funds to pay the project stakeholders. At this stage, different construction targets and deadlines need to be considered, and the potential impact of any foreseeable delay on the final budget is estimated
  - The process to develop a successful and cost-effective project is quite straightforward. However, during this process, any obstacle, misunderstanding, geo-political turmoil, or even weather disturbance can cause an unforeseen impact on the final project cost.

4.3.7.1 Some Recommendations When Starting the Work:

Early Stakeholder Engagement and Project Coordination:

- Usually, each stakeholder (e.g., prime design consultancy, sub-consultants, costing teams, means and methods team, etc.) is engaged at a different stage of the process. For instance, the consulting team that draws up the schematic design during the tender may have little or no overlap with the team engaged in drafting the detailed design. Since each entity tries to optimise its scope of work and profit, misunderstandings and conflicts may arise if incentives are not aligned and controlled.
- Some asset owners, such as railway operators, deploy an integrator: a key team member responsible for harmonising different stakeholders in the design and construction process, preventing clashes and pitfalls. In this regard, the agent responsible for achieving a proper cost-design project in construction is the person who performs the role of a structural integrator. This role leads the different stages of design, facilitates information transfer when stakeholders are engaged and disengaged, and tracks potential oversights that could arise. This role was traditionally carried out by the main designer, but since the advent of new or modern construction means to accelerate construction, the leadership for this role has become less clear. As the figure of a structural integrator is relatively new, its specified duties are not fully defined. Expertise in designing, a sound knowledge of construction techniques, and familiarity with the cost figures of the different agents involved are key for the performance of this role.
- In addition to adopting a holistic approach to bridge design, the early engagement of a structural integrator can enhance communication flows and improve continuity between different stakeholders in the design process. While using a holistic approach to optimisation and engaging a structural integrator does not guarantee the prevention of all rework, these two strategies can go a long way to mitigating the risk of costly amendments and oversights. Working together, they can significantly contribute to developing the next generation of cost-effective and innovative bridge structures.
- Another critical point to be highlighted concerns the modifications introduced during the design or construction process. Any change may affect the whole concept and lead to a repetition of tasks, resulting in subsequent impacts on cost and development time. To avoid design delays, changes of drawings, repetition of calculations, or redesigned structural shapes, it is crucial to involve all participating agents in the designing process, ensuring that the correct decisions are taken. For example, big gantries used to speed up construction are items of equipment that will need supporting from the bridge. If these high loads were not initially considered, a major change in the design will be faced, such as new dimensions, joints of temporary supports, braking points, or even temporary links when extreme weather conditions appear during construction. In this regard, a final bridge design requires early con-

sideration of construction and erection methods. Ideally, potential means and methods should be considered at the initial stages of the design process.

Bridge Technology Evolution:

- It is imperative to foresee these conflicts to prevent expensive rework. Bridge technology has traditionally aimed to minimise materials, but in current times, due to limited budgets and tight schedules, the deployment of different construction methods leads to different outcomes in terms of cost, schedules, and structural typology. Approaches like balanced cantilever construction or cable-stayed bridges, originally envisioned for cast-in-place solutions, now need to be slightly modified to accommodate prefabricated elements (e.g., segmental bridge construction). Prefabrication extends beyond reinforced concrete pieces; reinforcement cages can also be prefabricated. For example, when considering a movable scaffolding system (MSS), expediting the reinforcement placement is necessary to avoid delays in the advancement of the MSS. When implementing a mechanised system, the interaction between the main structure and temporary scaffolding needs to be analysed to establish which element will support the temporary loads before the completion of the entire structure.
- The total cost of a bridge should include all the expenses which will be necessary during the service life of a structure. Bridge costs fluctuate regularly due to evolving new construction systems. The new trend of accelerated bridge construction prioritises construction time and involves comprehensive designs considering various stakeholders in bridge design and construction beyond material costs alone, tending to yield the most cost-effective structures.

Temporary and Permanent Work Integration.

- Engineering trends are trying to split apart what is considered the main design and temporary works. However, these two stages cannot go separately, because the support of temporary works such as gantries, falsework, MSS, incremental launchers, etc. must be considered. If a bridge is considered just a balanced structure only when it is finished, any temporary stage not considered by temporary supports will affect the result since a redesign will be necessary to achieve safe and balanced temporary structural shapes. In this regard, a bridge cannot be imagined or designed without considering the construction methodology from the outset.
- Developing an optimally cost-effective bridge requires a multi-faceted approach and tight control to mitigate potential oversights and pitfalls. As mentioned above, the method deployed to erect a bridge's superstructure has a significant impact on the final cost of the project. It is vital to consider the intended method from the very inception of the design process. In addition, any unique load cases, or accommodations for temporary works (falsework, MSS, etc.) should be investigated at the earliest time possible.

To achieve the most cost-effective construction of a structure, several considerations need to be addressed. First, in the design stage, once the alignment is obtained, several configurations of span length, deck typology, and pier and foundation predesign must be calculated to be able to compare the materials of each solution. An analysis of figures with the ratio cost-to-span length will give the idea of what the best solution is for further development. When different chosen span lengths lead to specific construction methods, it becomes crucial to factor in the construction means and time within the cost-to-span length ratio, as they significantly influence the final cost.

4.4 Logistics: Critical Challenges in Bridge Construction

António Luíz Souza

Brazil / USA

4.4.1 Introduction

It is well known that with supply chain ecosystems going through an impacting transformation, it is important that logistics managers stay up to date with their delivery statuses and schedules to prevent bottlenecks, rework, and inefficient delivery processes [1].

It is essential to consider preliminary studies when identifying the project solution at its initial or conceptual phase, such as planning and logistics, all cohabitating in the same spectrum. This convergence of the analysis of work plans for projects located in remote areas is fundamental due to the natural and structural issues being more frequent and important.

This type of construction is present in urban and remote areas, and both have logistical analysis requirements, with coinciding and diverging sections of content. Among others, aspects such as:

- Characteristics of the project
- Geographical location
- Project size
- Material resources
- Construction time frame and delivery deadlines
- Interfaces and project delivery phasing
- Project funding
- Local conditions
- Interference and obstacles
- Specifications and regulations
- Construction means and methods
- Environmental, aesthetic, and legal issues
- Political and social issues
- Weather conditions.

These are examples and conditions that create parameters for better planning execution, where each must be considered for impact and composition in the project construction logistics.

4.4.2 Possibilities and Known Solutions

The varied reasons leading to the creation of a specific bridge construction project are not discussed here. However, from the outset, in the conceptual phase of engineering, logistic aspects can be identified to define the type of certain design elements, depending on the following aspects:

- Purpose – the nature of the use of the structure regarding whether road, rail, pedestrian, services, or all traffic is included
- Typology – the structural system considered for transposing, whether simple or continuous spans, arched,

suspension, suspended, or cable-stayed shapes and if concrete, steel, or mixed structures
- Geographic location – if the project is in an urban or remote area and the relief conditions, crossing of water flows, or vehicle traffic
- Environmental conditions – if the given solution covers and mitigates project impacts in design, construction, and operating phases, including any issues related to natural threats to its durability and maintenance
- Size – the characteristics and dimensions of the project regarding length and geometry
- Construction method – considers the construction sequencing, means, and systems to be adopted, as well as the source of materials and supplies available for construction.

These few elements mentioned above already determine the assessment for construction logistics in line with the builder's experience and capacity. Almost always, in the conceptual phase, most design-related project aspects have not yet been defined by the builder, and this task is delegated to experienced designers and consultants to study alternatives and feasibility with preliminary and limited information. This often results in ambiguous solutions that may motivate modifications at later stages of the project, with reflections on time and cost of execution.

Certainly, partially or combined, the factors mentioned above will make each project unique, making the logistics planning exercise for construction an important task to achieve the best performance objectives in execution.

A good example may be a hypothetical roadway overcrossing projected on a wide and flowing river located in the middle of the Amazon, designed mainly in a cast in place, multi-span, and cable-stayed concrete bridge, far away from the source of concrete materials, with limited river or air access and scant construction labour force. In addition to that, other challenges like hot weather conditions, transportation by barges, rip currents and seasonal high variations of river levels are to be expected, as well as ecosystem-related issues. Perhaps a better solution would be to design a suspension bridge with precast composite segmental box girders, with larger spans that permit the least number of piers over the watershed, able to transport precast and segmental girders, mitigating the transportation risks of using vessels or navigation traffic, and with lower resource use at the worksite.

Technological advances and mechanisation have also been incorporated on a daily basis into construction methodologies to increase productivity and benefits for cost reduction and execution time frame. In this context, processes of modularisation and standardisation have continued to improve projects' performances since their inception, providing better resource usage during construction.

4.4.3 Critical Issues

The construction process might be impacted for many reasons. Some common causes of frequent errors are as follows:

- Conceptual errors or direction in defining the structure typology
- Errors in resource evaluation for usage during construction as to the proper logistics associated with project location and market availability
- Lack of information or poor assessment of construction

- Errors in conceptual project planning, sometimes following project typology and design guidelines.

4.4.4 Project Location and Geography

Roads are elements of communication and flow of transportation means and may encounter natural obstacles along their route such as valleys and watercourses or also unlevelled intersections of vehicle or rail traffic, which are continued through bridges or viaducts. These structures may be located in zones of diverse characteristics that require varied solutions to accommodate design, aesthetics, construction, as well as environmental and legal requirements.

Construction logistics play a key role in the construction programme, where all resources associated with the working methods to be used during execution are considered. Each stage of construction requires components of labour, materials, and equipment in each subject area, the flow of which must be planned and executed precisely so that it occurs according to established standards and requirements. This combination of factors requires adopting a detailed work plan.

The layout of the production areas, the means and methods of accessing the work fronts, the transportation of materials, and the workforce skills are essential when drafting the work plan.

A river overcrossing project located in an urban area may require dual logistics for both land and water work fronts even though they have the same uses of material resources and similar methods. However, the means and methods may differ, thus partially affecting the concept of logistics, considering the use of vessels in transportation and as work platforms. In addition to the duality of the concepts mentioned above, bridge projects located in remote areas also require greater planning of staging areas and lodging to support construction execution. On land, interference with existing structures, services, and traffic routes is very common. In water, conflicts usually arise with the navigation and natural conditions of the environment.

4.4.4.1 Project Scope and Characteristics

Large-span bridges are renowned for the use of modulation in their construction process, regardless of whether it will be in concrete, steel, or mixed. Segmentation leads to the formulation of repeated processes, arrangements with prefabricated elements, operation of assembly, and movement of loads using equipment designed specifically for these activities. The natural sequence of the structure's action plan involves the execution of the infrastructure (foundations and pile caps), the meso-structure (columns and pier caps), the approach to the access road, and the superstructure (deck). This flow tends to occur with a lagged simultaneity pattern, according to the execution deadlines of each stage and the amount of work fronts associated with each subject area, limited by procedures and technical, economic, and operative aspects. The management of staging areas must be based on proper sizing to meet the needs of the necessary construction programme steps.

4.4.4.2 Specifications, Codes, and Regulations

Design and construction specifications and codes are highly important due to the need to make the work cycles longer or shorter to fulfil stages or sections of work to ensure the safety, quality, and durability of the works. The current requirements of the AASHTO LRFD standard embody one of the more comprehensive normative references for the design and construction of bridge and crossing structures.

Many other requirements set by standards influence construction processes and logistics.

4.4.4.3 Bridge Type and Purpose

Choosing an intersection structure design type is not usually a builder's assignment. However, already in the bidding phase, the builder faces enormous challenges in establishing the proper logistics and planning for the work execution. Examples include the construction of large bridges designed in remote areas, where sources of building materials, skilled labour forces, and access conditions are not well known to mobilise construction resources required to carry out the work as needed.

Low-profile overcrossings with modulated span structures, suspended or even cabled stays are the most common and most considered projects in concrete, steel, or composite materials. Alternatives to tray execution may range from cast-in-place to precast modular solutions with beams or box girders.

Each of these solutions will have an impact on logistics, the composite implementation of construction as well as for resource transportation and assembly of elements.

4.4.4.4 Project Time Frame

The project construction time frame is crucial to size work fronts and staging areas and how the work will be phased, to define cycles and execution steps. Challenges in figuring out work performance in shorter terms are increasingly frequent, mainly for economic and financial reasons, to meet the project return and deadlines. In addition, the progress of technology, construction processes, improvements, and competition among builders always matter in new projects.

Even so, when projects are in hard-to-reach areas, the obstacles and resource shortages drive the parameters in the logistics matrix and will be paramount in developing the master plan to achieve the project objectives.

4.4.4.5 Project Material, Construction Equipment, Supplies and Manpower

The choice of project type and construction materials involves not only the design structural solution but is also based on the actual conditions for the transposition and construction approach evaluated in the early stages of the technical and economic feasibility analysis.

Moreover, an early constructability analysis can determine the possibilities of equipment optimisation, or at least direct the builder towards the construction processes that could be adopted during the execution, to define sources of materials, like aggregates for concrete, or manufactured products such as cement, reinforcing steel, and other supplies.

The availability of labour on site is part of the initial investigation, as it may imply focusing on mobilisation and training prior to the start of the activities in which it will be applied. Sometimes, this task may be easily biased by socio-political and environmental issues.

4.4.4.6 Legal, Environmental, and Social-Cultural Aspects

Legal, aesthetic, environmental, and sociocultural aspects are considered in the project design and execution of a crossing. All of these have an impact on the logistics associated with execution planning. Facts such as expropriation of areas, migratory flow of animals, fishing, navigation, proximity of nature, or native reserves and strategic areas are examples of obstacles present in the development of the project and execution of works of this nature.

4.4.4.7 Project Management and Funding Plan

Funding sources are generally managed by project stakeholders and mostly move along the project phases. For the construction phase, an appropriate understanding of project cash flow should be accomplished and monitored to ensure the necessary acquisitions and progress throughout the project. Mobilisation and logistics implementation are expensive activities that involve advanced commitments among parties.

4.4.5 Good Practice in Logistic Analysis

Experience shows and it is well-known that logistics is one of the most important requirements in the workplan elaboration of any construction, especially in the case of bridge and viaduct construction projects, whether small or large. It is necessary to dedicate most of the time to the analysis of the components of the detailed work structure, to alternatives, or to task execution options, considering in each of them the means and methods to be applied and the source of its components, such as the workforce, materials, equipment, and support [2, 3].

Variations in complexity may come in different forms according to the various types of structures, their materials, dimensions, and the applied construction methods. The solution of this matrix makes each project almost unique. The builder's experience contributes enormously to the search for solutions that enable the construction to meet its schedule and the forecast of costs for the project.

Work in remote areas requires additional consideration of the factors of availability and accessibility of onsite resources, weather conditions, site relief, and the work environment.

Harnessing cutting-edge technology can lead to additional productivity gains in both administrative and production tasks.

As an example, when working with deep foundations, builders should factor in that a mechanic pile-driving rig will be required for generally driven piles, whether precast concrete or metal profile, or even auger cast, hollow drilled shafts, or concrete-filled, usually configured with a large diameter circular section. This construction operation, whether onshore or offshore, requires very detailed planning, execution, and controlling processes, with high intensity and application of resources. Skilled manpower is required, training, coordination, and precision in those operations, which are usually carried out by using heavy equipment. In offshore operations, other components related to navigation and vessel stability are added, which impact the performance of certain tasks.

The definition and choice of the right equipment and manpower to be used in those operations must comply with strict criteria and diligence to provide as many of these resources as possible at all stages, as in most cases they occur in confined areas where safety issues are commonly raised.

Comparisons between alternative operations should be evaluated to indicate the best choice to be taken as a priority and also consider some alternative plans for possible exposure to risks within a cause-and-effect study.

4.4.6 Fundamental Steps to Define a Logistics Plan

4.4.6.1 Understanding the Project Scope of Work

All projects, whether small or large, and especially overcrossing projects, require a detailed definition of all the infrastructure and logistics necessary for the main project execution, required means and methods, and preliminary activities for construction, according to the desired deadlines, targets, and goals. Furthermore, this includes the analysis of alternatives and solutions for construction processes, feasibility, complexity, risks involved, associated costs, as well as the appropriate use of resources according to origins and availabilities.

The process should start with the analysis of the project, regardless of whether it is at the conceptual, basic, or executive phase, looking at the construction phases and individual construction steps for each element. For example, in the case of a typical bridge with three spans, this would involve an enumeration of the phases for abutments, approach spans, and main span. The construction steps in each phase would be foundation, substructure (also called infrastructure), and superstructure. The definition of the construction methodology of each of the elements in each stage will require a description of its composition, such as the necessary material, equipment, supplies, and labour, according to the natural conditions to which the structure is exposed. In general, this type of work faces adverse conditions related to the geography, topography, and environment (land or water).

Looking at the activities that need to be executed on land, logistics should consider processes, routes, means and methods, supports, and equipment almost always at the ground level. When structures are to be built offshore, it is necessary to supply means throughout vessels (Fig. 4.13). The complexity of planning operations in both cases depends heavily on the dimensions and characteristics of the project and the pre-existing local conditions where it is placed. The guidelines for a work breakdown structure (WBS) elaboration for infrastructure and logistics do not differ from the usual process used in ordinary current projects, but it is important to ensure that the particularities of the activities are treated with the care they deserve.

The most relevant aspect in the course of the logistics evaluation process is to use experience, technical approaches, and alternatives to identify, perform, and accomplish the scope of work, according to the design and specifications requirements, interacting with the designers, specialists, consultants, together with the builders to find the best way to achieve results in terms of quality, time frame and cost for the project.

4.4.6.2 Due Diligence and Interactive Approach

At the present time, the elaboration of a 3D project and tools such as building information modelling (BIM) are being used to good advantage in design development, to perform virtual simulations and evaluate construction processes for better planning or even to adjust the design, with a view to better performance in construction and operation. This also fosters a better

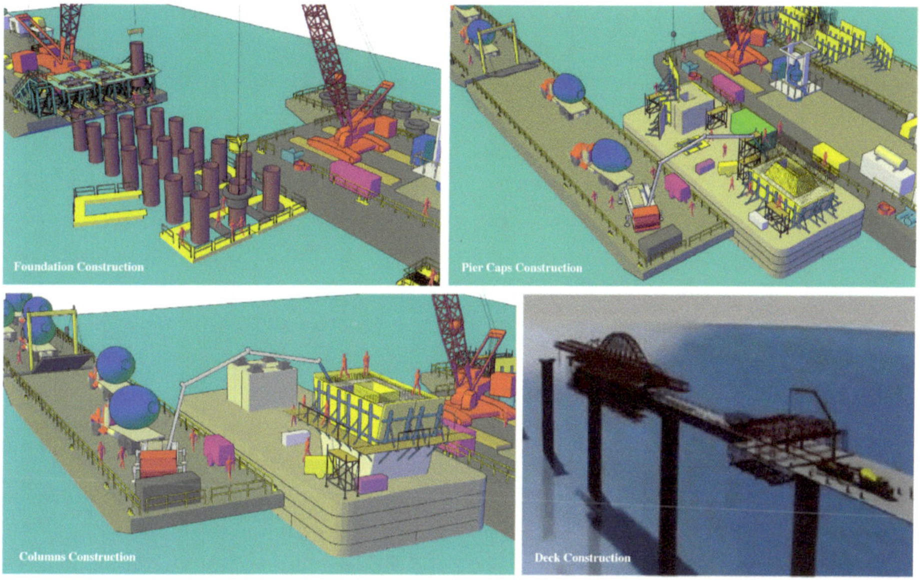

Fig. 4.13 Means, methods and infrastructure logistics for bridge construction

approach to the design, planning, and implementation of the infrastructure and logistics required for the construction stages of the work. Three-dimensional modelling allows the elaboration of animated sets for multidisciplinary analysis and interaction of activities, identification of interferences, and the preliminary analysis of alternatives and processes that promote modifications and accelerate the execution and safer management of the project. It is worth remembering these tools can also be configured in management models integrating planning and cost.

For bridge or viaduct construction projects in urban areas with high occupancy and mobility rates, 3D modelling and virtual simulations are fundamental to visualise solutions for physical interferences and public traffic management during the execution of the new structure. Construction logistics shows higher complexity and requires greater challenges among design and construction teams in figuring out alternatives that leverage proposed processes and resource availability to pursue the expected objectives.

Furthermore, the use of drones is already a reality starting from the preliminary phases and during construction such as surveys, implementation, images to the monitoring and update construction activities. Currently, scheduled and certified readings through QR codes are used to enhance control in fast-tracking, transparency, safety, and sustainability.

4.4.6.3 Development of Consistent Project Work Plan

There are well-defined processes to create a consistent work plan for construction infrastructure and logistics in project management and to control best practice which should be integrated with the master plan for the overall project construction. A suggested sequence is shown below merely for illustrative purposes:

- Understand project goals, objectives, and defined strategies and initiatives to carry out main tasks to be accomplished
- Gather the relevant information and past performance experience
- Establish team capabilities, roles, and responsibilities
- Estimate project timelines and budget
- Identify constraints and deadlines
- Set checklists, interaction, and review processes
- Evaluate results before going forward by phase.

4.4.6.4 Constructability Review

The constructability review process is defined, structured, and developed by the construction industry to provide in-advance planning, construction knowledge, and expertise to ensure design and implementation plan practicality and cost-effectiveness. The constructability approach primarily covers aspects related to project execution phasing, including detailed design scheduling, procurement, mobilisation, delivery of material, equipment and supplies, contracting, safety, temporary facilities and infrastructure needs, facilities construction, commissioning, and team organisation.

A constructability programme should be developed to outline and document the project policy regarding constructability. The plan should consider the minimum criteria to define constructability objectives, the general resource roles and responsibilities involved, and a feedback mechanism for contributions to project construction improvement, not limited to the following:

- Clearly defined and, where possible, quantified project objectives
- Constructability organisation chart defining roles and responsibilities
- Constructability plans and implementation guidelines
- Methodology for key performance indicator (KPI) measurement and verification
- Analysis of KPIs and interpretation to provide project performance benchmarking
- Procedures for capturing feedback and learning outcomes
- Constructability tools: forms, agendas, checklists, and workshops
- Issue and change resolution methodologies
- Targets and deadlines for constructability workshops.

4.4.6.5 Logistics Risk Matrix & Challenges

Regarding the matrix of risks and challenges proposed for the particularities of infrastructure and logistics construction works, the methodology does not differ from the main work and in in general converges in the similarities and differences of the activities which make them unique. Following the procedures and processes adopted by the management system, it is not expected that divergent results should lead to the adoption of a specific policy for this.

Constraints:

- Design criteria
- Legal and environmental provisions
- Limited information regarding the availability of space for the implementation of support capacities in urban areas
- Mobility traffic management in urban areas

- Maritime traffic in navigable water projects
- Natural geographical and climatic conditions
- Availability of material resources, equipment, and the workforce

Critical Issues:

- Design review and approval process;
- Land acquisition and liberation;
- Environmental assessment.
- Public relationship, communication.

4.4.7 Lessons Learnt and Recommendations

- Logistics management is directly and indirectly related to productivity loss, and logistics can be the sole reason for a major project time failure.
- It is fundamental that the logistic context of a bridge is duly considered and evaluated from an early stage, i.e., right from the design phase, or even from the definition of the terms of reference.
- The logistic context may have a relevant impact on the ecological footprint of a bridge, particularly on CO_2 emissions.
- The logistic context may change, sometimes dramatically, with profound impacts on a construction process. This knowledge should feed into a prudent bridge construction decision-making process.
- With supply chain ecosystems going through a transformation of their own, it is important that logistics managers keep up to date with their delivery statuses and schedules to avoid bottlenecks, rework, and inefficient delivery processes, where a major aspect of value is flexibility.

References

[1] A. Ekeskära, M. Havenvid, T. K. Gustavsson, P. E. Eriksson, Construction logistics in a multi project context: coopetition among main contractors and the role of third-party logistics providers, Construction Management and Economics, Volume 40, 2022 – Issue 12

[2] F. Wang, Lihuan Ma, W. Huang, J. Yan, The Study of Logistics Management Theory in Material Cost Control, in International Conference on Wireless Communications, Networking and Mobile Computing, 2008

[3] Peter Seitz: Increasing Planning Quality for a Project, 40th IABSE Symposium Nantes, 2018

4.5 Scheduling Different Construction Methods

Thierry Duclos

France

4.5.1 Introduction

Scheduling different construction methods is a major issue in terms of tracking the time consumption in which each actor of every construction act is involved.

Civil works are not the result of only one actor. On the contrary, they are the result of numerous actors, and obviously, scheduling is a management act that aims to minimise duration.

It is crucial to look for the relevant information from previous experiences and the best implementation method for building bridges. Finding them requires considering significant sources from which key information for this purpose should be extracted.

4.5.1.1 Bridge Construction Method Choices

Building a bridge requires a construction method drafted during the design period.

This method will define a planning process giving rise to a schedule [1].

The most appropriate method draws on the results of several analyses in which there is the major item (the bridge) and also secondary items that can strongly affect the progress of the works.

For example:

- Environmental impact limitations can involve typical methods: an incremental launching method could be preferred to a cantilever method due to a lower risk of pollution onsite
- A precast bridge where the length of the deck allows significant savings compared to a classical cast in situ
- The area for work installation may be a significant constraint for the adopted method
- Major events during construction: for example, flooding of the site
- The owner can impose constraints such as pier positions that exclude one solution over another
- Risk analysis can define complementary goals to secure the work.

The database must be constituted to compile the relevant information for the planned method.

This preliminary approach shows that the work process must be known. It will define a preliminary schedule incorporating several goals. Therefore, analysis is necessary to hierarchise these goals and determine their positioning in the overall process and the time consumption of each goal with their necessary impact level for a successful outcome.

The builder (engineers, design team) must have in-depth knowledge of bridge-

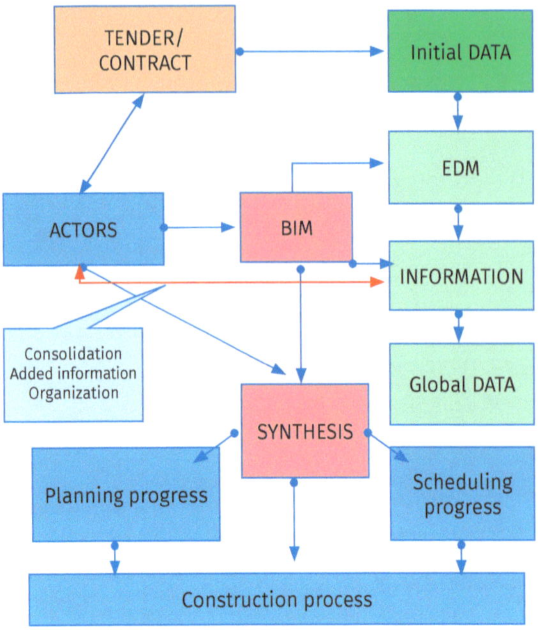

Fig. 4.14 Internal link of an operation (EDM: electronic document management)

construction methods and works planning. A simplified diagram with the internal links of an operation is presented in Fig. 4.14, and a similar drawing can be provided with the external relationships having an impact on the works.

This links diagram is also a way to obtain savings and more integrated work and design.

4.5.1.2 Methods Partial Decisions

Building a bridge is a comprehensive management issue.

But this is not completely true as specific elements of the bridges like foundations, piers, and the equipment to install require their own planning approach. So, a secondary analysis is necessary to complete the first overall analysis mentioned above. Each component will have its own impact and its own constraints.

Similarly, for these secondary components, the most appropriate method for each needs to be defined and analysed.

The overall schedule will be completed by different deadlines depending on the planning of the works and the schedules of each component, each with its own constraints.

4.5.1.3 Works Organisation

The organisation of the works is defined from the planning process, requiring teams and actors.

While short bridges can be considered very simple civil construction works, major civil works need multiple actors such as site workers, engineers, method engineers, accounting managers and so on.

The essential point is to organise this team which, like the many components of a bridge, includes numerous groups of people, each with their own aims and skills.

These groups begin their actions at specific times. Therefore, the broken-down segments of works must be arranged to prevent collision in time or space, waste of time, and inappropriate action onsite, especially if these segments are on the critical path.

Schedule management is necessary to structure the actions of each party involved, but also to obtain overall coordination of all areas to identify non-compliances and clashes [2].

4.5.1.4 Unknown Events

It is a fact that random events create delays and often disorganisation, meaning that the schedule cannot be respected, where some of the underlying reasons are:

- Incomplete risk analysis;
- Lack of procedures due to inept description, poor knowledge of

technical details, or misevaluation of the means;
- Insufficient investigation leading to lack of information, e.g., with a possible potential risk of undervaluation of the foundation strength due to an economic restriction of the ground survey;
- Unsuitable data due to a vagueness of the technical requests.

The initial database should be detailed as far as possible in the main domains required for the construction.

4.5.1.5 Desing Team Involvement

As explained in [1] on improving the design process, early contractor involvement and integrated project delivery both have a positive impact. These approaches value participatory methodologies. The tendency is to integrate all parties: the owner's team, the designer's team, and the contractor's team, to produce a specific team for a defined construction site.

This tendency implies the allocation of tasks, expenses, responsibilities, and commitments. This sets the scene for cooperative work in which different hypotheses can be studied where nothing is done under threat of being purely gratuitous.

This is the best way to involve the designer throughout the entire project, particularly in the method construction process. It simplifies any adaptation of the bridge layout, if necessary, and the owner's association may reduce the time to make an informed decision.

In the past, the designer was frequently not fully involved in the construction. In the end, the owner and contractor alone decided on the last processes for the construction, meaning that there was a lack of the designer's engagement in subsequent stages. Often improvement possibilities were limited due to the fixed design with no freedom for adaptations: the calculation notes had already been done, the designer was no longer engaged, and so forth.

Accordingly, it is evident that the works contract must be adapted to include the designer throughout the entire duration of the works, and this fact is explained to the contractor before he starts.

4.5.1.6 Information Management – Workflow

Poor information management is an issue with immediate consequences on the schedule. A workflow must be planned, shared, and checked for requests, clarification, orders, and general information.

In the first stage, the methods management should find the means to optimise the work training, incorporating information about weather, major events, and so forth, so that the chosen methodology integrates the actions that will limit the impacts. This workflow will help the contractor to optimise the schedule, human resources, and materials means. The time frame to engage in this action and these means must be implemented in the general schedule.

4.5.1.7 New Technologies and Implementation Time Consumption

New digital technologies endow a smart approach to drawing up efficient schedules. These technologies include, for example: building information modelling (BIM), scheduling software, planning software, and document management software (EDM).

According to the extended capabilities of the software, different parts of the work process will be integrated into the data. This requires obtaining relevant information, its verification, and incorporation into the works schedule.

In this analysis, the main checkpoints should be identified. These points are critical moments where the checking of work processes shows possible delays and other issues.

The time implementation of these tools must also be integrated into the schedule and is one of the important actions considered in the planning.

4.5.2 Possibilities – Known Solutions

4.5.2.1 Document Management

The sections presented so far point to some foregone conclusions.

Firstly, the personnel must be hired, prioritising knowledge, skills, understanding, and interests. [3]

Tools will be used to manage and to share information, evolution, and orders, including:

- A virtual platform which should be active as soon as necessary
- Communication tools such as the electronic document management (EDM) system;
- Building (bridge) information model may be used (especially in urban zones where a lot of information must be integrated)
- Creation of collaborative platforms with links to EDM. These collaborative platforms may be open spaces with joint actors: designer, architect, contractor, owner, or management committee to drive the collaborative platform, check the relevant information, and for decision-making in meetings
- A technical committee to manage technical issues and potential risks
- Security management to guide safety in design and work safety.

These tools enable the project team to manage the significant quantity of generated information: drawings, minutes of meetings, and procedures.

The information management will need efficient managers to obtain useful information, with the following outputs:

- Complete definition of the work allocation
- Evaluation of energy consumption with time monitoring, assessment, and statement
- Consolidation of the schedule.

The actual organisation of electronic document management can be illustrated as in Fig. 4.15:

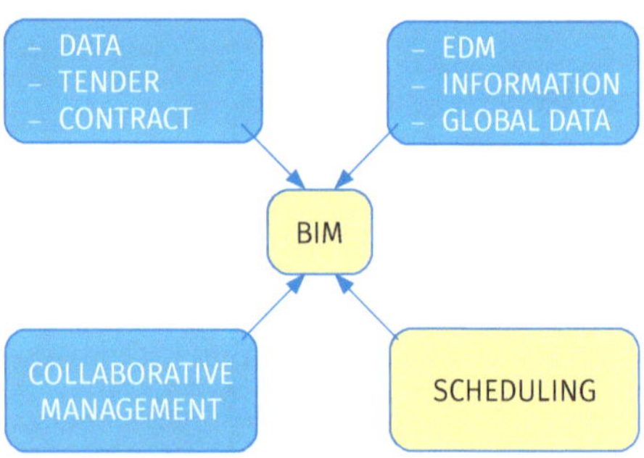

Fig. 4.15 Organisation of electronic document management (EDM)

Further information can be found in [4], with the project organisation being as described in [5], and visual methods presenting all aspects of the works with integrated tools.

4.5.3 Critical Issues

4.5.3.1 Information Management

Collaborative management and collaborative organisation must be initiated at the beginning of the contract.

This needs time, space, tools, and humans with skills. This space is often created between the construction actors managed by the main contractor in the form of a collaboration platform. This platform can be linked to building information modelling (BIM) or electronic document management (EDM).

The choice of the BIM or EDM and initiation of information management requires clear identification of the necessary specifications, based on the following key information:

- An agreed list of documents
- The approval circuit with agreed terms
- The definition of a clear numbering system
- The production schedule of each actor
- The monitoring of actors' observations
- As well as other specific details depending on the project.

Although the management system of work has evolved enormously in recent years, the implementation of different systems is still not fully refined, and in most cases, it requires a tailor-made solution. This aspect is frequently traced back to the many associations of actors in different management systems and distinct cultures, which often cause rework and delays.

Other challenge arises from the building information modelling (BIM), the structural model, electronic document management (EDM), the schedule software and so forth, because the main issue is the inoperability of this plethora of tools due to lack of standards and the difficulties of using data in different configurations [1, 6].

Data security is also one of the main objectives to improve the situation. The data must be consolidated and kept safe, with the main risk being data corruption due to misunderstanding or poor rankings. In brief, if data is not safe and well organized, with adequate understanding of contents, for example, adequate parametrization, data may be inadequate for its purpose.

Contributors must be competent on a wide skill spectrum,and the elaboration of the building information modelling (BIM) collaborative platforms needs a common data environment (CDE) which will bring:

- Useful exchange of knowledge
- An accurate real-time update of the available information
- Reduction of errors
- Overall process improvement.

4.5.3.2 Lack of Answers in Certain Domains

As seen above, the inoperability of different software requires other software for data integration and exchange, entailing a wastage of time to implement them but also errors or omissions.

In this regard, some providers try to establish standards to improve this situation and limit the work involved in entering the data into other tools. However, efficient interoperability still has a long way to go.

Also, collaborative platforms are evolving, with the trend being to create comprehensive software in which the structural analysis, the building information modelling (BIM), the electronic document management (EDM), and other processes are interlinked. However, due to the costs of this type of approach and the risks involved in obtaining an appropriate tool, progress is currently slow. This means that efficiency remains in the hands of competent personnel, whose skills are analysed and described, although their efficiencies are not apparently replicable. So, progress is clearly necessary here.

4.5.3.3 Risk Analysis

The development of project management has defined a typology for risk analysis, and the experiences and exchange of experiences have produced extensive data that needs to be used and transferred to new projects.

- **Advisory Committee**

At a time of increasing fake news, many studies have developed knowledge about thinking bias. Academic knowledge and research can be polluted by cognitive bias. Certainties learnt during studies or imprecise information can become obstacles or bad source options. Engineers and managers have not been vaccinated against these abuses. A simple solution is to draft an overview with a scientific committee giving opinions on significant options for the project, where the staff involved should be skilled and flexible.

- **Feedback**

All types of construction methods are now thoroughly researched, and their planning is well known. Many articles have been written, supplemented with data drawn from experience.

- **Random events**

Climatic events will always be random to some extent, which is difficult to eliminate or mitigate. However, these events are related and make it possible to draw lessons for construction. As these events are often newsworthy, libraries are excellent places to compile relevant information. Integrating all these aspects in a brief report gives tools to managers, designers, and consultants.

However, many obstacles can hamper this risk management:

- **Inappropriate standards**

It is a common situation in construction to stumble across unexpected provisions in the contractual standards. Therefore, it is necessary to have options to solve this issue with processes agreed with the owner or independent reviewer.

- **Individual lack of knowledge about risks**

Knowledge about risks and their management are skills that could bring savings, but also a better evolution of the planning. Hence, it is important to schedule training in this field of expertise.

- **Lack of know-how**

The staff must be coached by experienced teachers or experts with their own experience.

4.5.3.4 Cost Impacts

The cost-effectiveness analysis of construction methods aims to limit quantities but also to optimise organisation. Considering all the aspects presented above is a step in the right direction towards limiting waste and throwing out inappropriate elements or designs.

Each user feedback must be well-documented and formalised. Taking feedback into consideration is crucial when creating historical data (time series) to produce a cost history with expected gains and losses.

4.5.4 Good Practice

4.5.4.1 Traceability, Digital Implementation, and Experience Feedback

It is necessary to develop an experience feedback library and digital information, providing a summary of experiences. This requires experts to compile extensive research and set up a data library listing all the significant technical points.

As explained in [7], digital tools will carry out more precise engineering which will be completed with summarised experience and knowledge, providing for a more consolidated project. This compilation brings in new knowledge with savings.

4.5.4.2 Choice and Follow-up

This paragraph is a reminder that all technological and managerial choices must be followed up with a documented history, gathering events and evolutions, explaining the positive and negative aspects during the development of the process. This work is often done at the end after completion of the project, so relevant information is lost.

4.5.4.3 New Technology

One of the solutions is to limit the number of software applications that are used. This strategy aims to improve risk identification, implementation, randomness, and experience feedback. It must be accepted at the beginning and shared among all the participants. It can involve new costs to update digital capabilities.

Another option is to provide a collaborative operating platform. Recent examples show gains in terms of cost and time, such as the case study presented in the reference [8].

4.5.5 Trends

The use of the integrative and parametric methods [4, 9] adapted to the integrated team with a collaborative platform leads to the dynamic elaboration of construction methods to cut cost and time but also to increase safety. Just as progress has been seen in software capabilities, the use of REX data is among the trends seen in construction firms and design teams.

4.5.5.1 Impact of New Technologies

With the development of building information modelling (BIM), many teams developing software aim to interlink all the potential models required in the projects. This possibility is becoming affordable due to the evolution of consulting team skills and the rapid progress of artificial intelligence software able to analyse the steps in project preparation. These developments could change the paradigm of project preparation as this has already been incorporated by engineering schools and universities, and new jobs are certainly going to emerge.

4.5.5.2 3D Printing

This technology can be used to replace staff for specific parts of the construction: repetitive elements, trivial technology, etc.

It can be a means of saving time and costs with clear impacts on safety and can also be incorporated in developing strategy. For construction planning and the complexity of site operation, there are opportunities to check and analyse the robustness of a specific solution with a preliminary simplified version drawn up through 3D printing. It is imaginable that artificial intelligence can be used to test construction proposal profitability through a 3D printing scenario.

4.5.5.3 Deep Learning

Deep learning is a technology that helps to develop efficient software working with intelligence, adapted to repetitive actions done by robots. Control and management tools are areas where this technology is applied, but it can be extended to the overall construction process.

Automation in civil works represents the current trend of evolution, in which dreams of the past have now become a reality. This technology needs skilled staff and new jobs.

Many construction methods involve repetitive actions which can be integrated into a software management process. Indeed, technical solutions are already used to gain savings, productivity, and safety in repetitive methods of construction [10, 11]. To achieve automated tools, machine learning is the first step, requiring data with relevant information on work methods [12]. This mechanical approach can be extended to management methods. For example, in scaffolding construction several methods can be used, and basic data could be analysed with a view to saving materials or cutting costs, scheduling, work sequences, etc. This approach could be applied to other methods using repetitive actions and these analysis models could be used during work optimisation as well as during tenders.

Extending this approach using limited skilled staff and automated tools could drive an overall project, where the assistance of deep learning is the next step forward from machine learning.

4.5.5.4 Automation and Safety

With this integrated approach using automation and including safety in the design, current projects will be safer. Moreover, these methods of working, deep learning, comprehensive data, and experience feedback analysis data will lead to unprecedented progress.

But is there an asymptote? With the increase in computer capacity, methods, planning, scheduling, and management it will definitely progress, opening up a very interesting future with new roles and opportunities.

4.5.5.5 Conclusions

The civil works domain has not followed the same development as the industrial domain. Its pace has been slower. However, for several years now, with digitalisation expansion in all domains of life, comprehensive data has begun to invade this domain, helped by graphic software, digital design, and building information modelling (BIM). Modernisation has come at a time when resource management is rising to a significant challenge, meaning that progress is moving in the same direction.

4.5.6 Lessons Learnt and Recommendations

Summary of previous good practice.

- To achieve reliable planning, it is critical to develop an experience feedback library and digital information, producing a synthesis of experiences;
- The use of integrative and parametric methods adapted to the integrated team with a collaborative platform leads the path to the dynamic elaboration of construction methods in order to cut costs and time but also increase safety; Knowledge about risks and their management are skills that could bring savings but also a better evolution of the planning. Hence, it is important to schedule training in this field of expertise.

References

[1] Peter Seitz: Increasing Planning Quality for a Project, 40th IABSE Symposium Nantes, 2018

[2] Brian Duguid and coauthors: The Ordsall Chord, Manchester, UK – digital delivery of design, 39th IABSE Symposium, 2017

[3] Mats Karisson and co-authors: Methodology for selection of production method in an early stage – improved conceptual design process, 39th IABSE Symposium Vancouver, 2017

[4] Rasmus Rempling and coauthors: Aspect of Integrating Design of structures, Civil Engineering and Architecture, 2015

[5] Peter Seitz: Visual Project Organization, IABSE Conference Geneva, 2015

[6] M. Maffeis and coauthors: One Single Model: a New Parametric Approach to Megastructures, 40th IABSE Symposium, 2018

[7] Jon Leach: Adding Value through innovation in Structural Design 39th IABSE Symposium, 2017

[8] T. Duclos and co-authors: A Particular Studies Management for an Unusual Bridge Project at Sea at la Réunion Island, Proceedings 40th IABSE Symposium, 2018

[9] Philippe Menétrey: The Integrative Design Method for Bridges, SEI, 2018

[10] P. Pacheco and coauthors: High Productivity in Bridge Construction - the OPS Effect, 2015

[11] P. Pacheco, J. Munoz-Rojas and coauthors: An innovative system of precast segmental span-by -span construction for span length of above 100m, Multi-span Large Bridges, Taylor and Francis Group London, 2015

[12] Serge Abiteboul -and coauthors: "What is AI?", Pour la Science, Mai-Juin 2022, HS n° 115.

5

Construction Technology

5.1 Specifications of Deck Construction Equipment – General Overview

Pedro Borges
Portugal

Very often, the deck construction equipment for a bridge is specific to the project, acting as a prototype either entirely or partly. Therefore, its specifications are crucial for the project's success. This section aims to provide practical information for the development of specifications for deck construction equipment. This is done by identifying the pertinent fields of knowledge, suggesting a possible specifications index, and pinpointing aspects that are commonly overlooked.

In the context of this section, specifications refer to all the prescriptions contained in the procurement documents (Requests for Quotation - RFQ) [1]. The content presented herein is directed at tailored equipment, such as movable scaffolding systems or launching gantries, and not for equipment that can be selected from a usage/capacity catalogue, such as cranes.

Equipment specifications embody the foundations of both procurement and supply. They are usually part of a contract that defines the equipment and associated services. Given that deck construction equipment knowledge is a highly specialised field of knowledge, it is crucial that the specifications should be developed by experts in equipment engineering.

Traditionally, deck construction equipment specifications can be categorised into two types [2]:

Prescriptive	Performance
Saying exactly what and how	**Saying how well the work should be performed**
A prescriptive specification may completely define an equipment (or a part of it). This type of specification focuses on the equipment solution itself, rather than the result/outcome of the equipment's work.	A performance specification defines the equipment (or a part of it) based on the measurable result/outcomes of the equipment's work. This type of specification is focused on the results of the equipment's work. It outlines what needs to be delivered but does not detail the specific equipment solution.

Equipment specifications are frequently composed of both types, addressing different aspects of the equipment. From the purchaser's point of view, various aspects can be assessed to identify the advantages and disadvantages of both prescriptive and performance specifications. Ultimately, the purchaser may implement strategies

<table>
<tr><th rowspan="2">Topics</th><th>Prescriptive (equipment solution)</th><th>Performance (equipment work result)</th><th rowspan="2">Notes</th></tr>
<tr><th colspan="2">tendency</th></tr>
<tr><td>Existent equipment allowance</td><td rowspan="5">Low</td><td rowspan="5">High</td><td>Performance specifications increase the likelihood of using existing equipment (already used in other works).</td></tr>
<tr><td>Innovation & alternative solutions allowance</td><td>Specifying the required work result can lead to innovative and alternative solutions. Investigation and deeper knowledge are traditionally on the supplier's side.</td></tr>
<tr><td>Supplier Engineering capability</td><td>From one perspective, if the equipment solution is specified, the supplier may simply be executing instructions.</td></tr>
<tr><td>Solution Risk transfer to supplier</td><td>Specifying the required work result transfers the risk of the equipment solution to the supplier. Specifying the equipment solution places the overall risk on the contractor, although responsibility for the equipment itself may shift to the supplier through contract clauses (this can be ambiguous).</td></tr>
<tr><td>Equipment price competitiveness</td><td>A specified solution may limit the use of standard mechanical components (e.g. hydraulics, hoists). It might not consider each supplier's specific knowledge and strengths.
Specifying the equipment solution could lead to engagement with lower qualification/ cheaper suppliers. However, this may be a contradiction since a purchaser specifying the solution is often more demanding in terms of quality.</td></tr>
<tr><td>Solution oriented towards the experience/ preferences of the contractor, operation crew and bridge designer</td><td>High</td><td>Low</td><td>Prescriptive specifications allow solutions to align with the past experiences and preferences of the contractor, operation crew, and bridge designer.</td></tr>
</table>

to eliminate or mitigate the disadvantages associated with each type of specification.

In an ideal scenario, the specification of an item of equipment should be the result of an overall optimisation process, not only focused on the separate parts (equipment, bridge, or operation) but also on the final result to be attained as an integrated process. To achieve this, the main players in the process: contractor, bridge designer, specialised operation crew, equipment designer, and supplier must all work together. The work should be mediated in a way to make the players fit together, optimising the technical solution, construction time, and total cost, and achieving a risk evaluation compatible with the contractor's profile.

The equipment specifications may be produced by the contractor, by the supplier, or by both [3]. The contractor may, at the first stage, produce a performance specification and ask the supplier for a corresponding prescriptive specification, in exchange.

A response to a Request for Quotation (RFQ) implies a significant investment on the part of the suppliers and thus, in an ideal scenario should be paid or imply a bonus according to the offer/solution classification. This could improve the quality of the response of the suppliers and allows bringing suppliers on board at an early stage of the project, which could possibly be the most efficient way of working.

The content of an equipment specification should cover various requirements which are categorised into 8 fields of knowledge in Fig. 5.1.

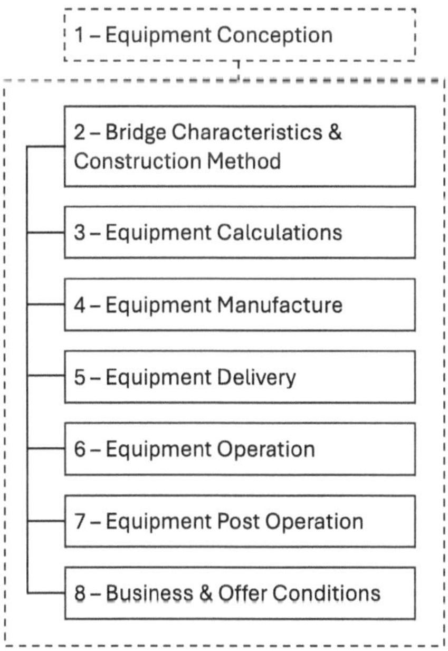

Fig. 5.1 Major knowledge fields for equipment design

The mutual interaction between fields 2 to 8 is intricate. Specifications and decisions made in one field can significantly impact the others, for instance, when considering the options of a tailored steel truss divided into large modules or a steel structure segmented in individual bars to fit into containers. The first option reduces calculation hours, fabrication hours, and significantly reduces assembly hours (positively impacting fields 3, 4, 6, and consequently 1), but it negatively affects transportation, requiring special road transportation or break-bulk (adversely impacting field 5 and consequently 1). The second option produces the opposite effect. The equipment design knowledge field encompasses all specifications from fields 2 to 8, conducts necessary analyses, and makes informed and weighted decisions. This method allows for the progressive development of the equipment solution.

5.1.1 Suggestion for Equipment Specification Index

An index for equipment specifications is suggested in this section. Creating such extensive and exhaustive equipment specifications may be challenging due to several factors: the unavailability of information, the involvement of multiple specialities, and the niche knowledge required. This suggested index can serve as a useful checklist to facilitate interaction between purchasers and suppliers throughout all stages of the business transaction. The depth of content in each chapter and the type of specification should be determined by the contractor.

#	Chapter suggested name	Suggested content
0	**Introduction**	
0.1	Project and scope	
0.2	Contractor policies	Code of conduct, safety, environmental, procedural, others.
0.3	Glossary	
1	**Equipment design**	
1.1	About the equipment design	Introduction, requiring the equipment to be designed to perform what is defined in all the other chapters. General policies and guidelines. Priorities (delivery schedule, solution, price). Any special/specific characteristics not explicit in the following chapters. Degree of mechanisation/automation versus manual operations. Degree of monitoring by sensors. Other related development aspects.
1.3	Regulations and certification	Regulations the equipment or parts must comply with (e.g. machine directive, CE marking, UKCA marking, European regulations). 3rd party Notified Body for certification.
1.4	Local plant design registration / local certification	Requirements for local certification or registration.
1.5	Equipment technical solution	Definition of the technical solution (e.g. geometrical definition, type of electro-mechanical systems, velocities).
2	**Bridge characteristics & Construction method**	
2.1	Bridge geometrical characterisation	Full geometry of the bridge and its surroundings.
2.2	Bridge construction	
2.2.1	Construction method	Definition of the method, kinematics, method of segment delivery
2.2.2	Scope for the equipment	Deck to be erected.
2.2.3	Construction sequence and stages	
2.2.4	Construction duration	
2.2.5	Number and type of equipment sets	
2.2.6	Interface points and loads	Expected interface points between equipment and bridge and corresponding expected loads, stiffness.
2.3	Bridge site space-time restrictions	Access to work fronts, working season, clearances from power lines, etc.
3	**Equipment design and verifications**	
3.1	Structural configurations	Identification of all structural configurations.
3.2	Standards, Norms, and Directives	Structural, electrical, mechanical, hydraulic, etc.
3.3	Combinations, coefficients, class of consequence	Coefficients adjusted, based on how well the load is known (e.g. deck segment weight).

#	Chapter suggested name	Suggested content
3.4	Actions	Self-weights, bridge loads, thermal variations, operational and out-of-service wind, seismic loads.
3.5	Operational temperatures	Defining electro-mechanical-hydraulic limits and steel sub-grade.
3.6	Type of connections	E.g. bolted, pinned, high-strength bars with shear pin.
3.7	Equipment life span	Structural, mechanical systems (loading cycles).
3.8	Calculations contractor and/or third parties check	Methods and duration for approvals by contractor and 3rd parties (phases, packages, loops).
3.8.1	Calculations 3rd parties check	
3.8.2	Professional engineer stamp (local)	
3.9	Equipment calculations insurance	
3.10	Other specific topics	
4	**Equipment Manufacture**	
4.1	Manufacture specifications	
4.1.1	Standards	
4.1.2	Material traceability	
4.1.3	Blank/Trial assembly	
4.1.4	Quality requirements	
4.1.5	Material grade and sub-grade	Referring to the applicable standard.
4.2	Quality Assurance & Quality control	
4.3	Manufacturing 3rd parties check	
4.4	Corrosion protection	
4.5	Warranty	Duration and conditions.
4.6	Manufacturing insurance	
4.7	Factory Acceptance Tests	
5	**Equipment Delivery**	
5.1	Incoterms delivery	
5.2	Type of containers	
5.3	Packaging	Degree of component split, maximum dimensions, and weight.
5.4	Packing lists and shipping documentation	
5.5	Transportation insurance	
6	**Equipment operation**	Tasks to be accomplished at the construction site
6.1	Assembly	Restrictions, sequence, provisional supports, partial assemblies.
6.2	Relocations and reconfigurations	
6.3	Disassembly	
6.4	Typical kinematics for deck erection	Sequences, preferences, drawings.
6.5	Special kinematic cases for deck erection	

#	Chapter suggested name	Suggested content
6.6	Elements to be handled by the equipment	E.g., rebar cages, formwork, post-tensioning hydraulics.
6.7	Accesses	Including emergency evacuation planning.
6.8	Safety	Standards, lifelines.
6.9	Electric network plugs & lighting	
6.10	Target cycle or task duration	Expected cycle duration and assumptions (e.g., shifts, working days, crew sizing, holidays, expected task durations).
6.11	Maintenance and periodic inspections	Type of maintenance permitted/required (e.g., manual versus automatic greasing).
6.12	Onsite acceptance tests	Part of the tests may have to be performed during the 1st span erection, confirmation of weight of equipment, reactions; Load testing criteria.
6.13	Commissioning & certification	
6.14	Assembly and site tests 3rd party check	
6.15	Crew training	Associated service.
6.16	Specialised crew	Team size and specialties.
6.17	Rules for equipment correction on site	Corrections for manufacturing errors and malfunctioning.
6.18	Rules for equipment modification on site	Improvements, adaptations.
6.19	Supplier support during construction	In case of breakdowns diagnosis and repairs. In case of additional training. Clarification of doubts. New issues.
7	**Equipment Post operation**	
7.1	Future applications	Increased operational limits compared to the first project. Easy geometrical adaptation to other bridges.
7.2	Equipment storage	General requirements for the equipment storage, if any.
8	**Business & Offer conditions**	
8.1	Supplier qualification	E.g., experience; reputation; accidents; Support during the whole project; manufacturing locations (quality & political risks); material origin; in-house works; subcontracting limitations.
8.2	Offers evaluation criteria	
8.3	Schedule duration (from procurement to bridge deck erection)	Request for Quotation (RFQ) release date. Meeting and clarification dates. Offers delivery date. Award date. Contract signing date. Equipment Delivery date. Equipment assembly starting date. Deck erection starting date. Deck erection finishing date.
8.4	Business Type	Sales, rental, buyback.
8.5	Scope: Base Offer	Detailed inclusions and exclusions matrix.
8.6	Scope: Optional Supplies	Detailed inclusions and exclusions matrix.

#	Chapter suggested name	Suggested content
8.7	Scope: Documentation of the equipment	Equipment programme and calculations assumptions. General drawings. Kinematic drawings. Calculation report. Assembly instructions. Relocation instructions. Operation manual. Disassembly instructions. Method statements & checklists for all works and operations. Maintenance manual. Manufacturing documentation (drawings, as-built drawings, quality process). Training material Risk analysis. Interaction with bridge (loads, interfaces, anchors, precamber). Storage instructions manual (rules for storage, rules for history record & documentation preservation, rules for maintenance and periodic inspections, procedures for future applications).
8.8	Payment conditions	Payment schedule, performance bonds, bank guarantees, prices, penalties, liquidated damages.
8.9	Intellectual and Industrial property	
8.10	Offer required validity	Validity date. Price correction indexes.
8.11	Offer deliverables	Clear definition of minimum obligatory offer contents. Valued additional offer contents. Supplier qualification pack. Information to allow solution analysis: Equipment drawings and description; Characteristics of electro-mechanical components (e.g. type system, velocities); Detailed Schedule Duration by Supplier (needed info to start, calculation stages and revision loops, calculation assumptions, targets and deadlines, material procurement durations, manufacture, factory tests, transportation, assembly, site tests, commissioning. Equipment working cycle tasks durations, e.g., launching. Operating costs: consumables, spare parts, energy sources and consumption Information to allow offers comparison: Inclusions and exclusions list. Price. Bill of Quantities (BOQ). All other information that contractor defines as important for the solution and offer analysis.

In the scenario of a prescriptive specification, Chapter 1 of the Specifications may take precedence over the remaining chapters. Conversely, in the case of a performance specification, Chapter 1 will be less detailed, and Section 1.5 might not even be necessary.

5.1.2 Lessons Learnt and Recommendations

Critical aspects of the documental organisation and requirement specifications are often overlooked. Enhancing these elements can significantly improve the quality of specifications and reduce unproductive work for all involved parties.

To improve the specifications documentation package and the quality of supplier offers, the following suggestions can be implemented:

- Ensure careful organisation of specification documents, which includes implementing an appropriate table of contents for logical and fluid reading
- Group requirements by fields to streamline understanding
- Concentrate the scope of work in an inclusion/exclusion matrix, which is more effective than dispersing information throughout the documents
- Avoid repetition in the documents:
  - Repetition can be more time-consuming for understanding and reviewing documents
  - Repetition often results in errors due to inconsistent statements
  - For multiple equipment types on the same bridge (even with different suppliers), it could be more efficient to create a single document for common information and separate documents for individual equipment specifics
- Eliminate unclear or ambiguous statements
- Clearly identify what is mandatory and what may be subject to supplier adaptation or improvement
- Use abbreviations or acronyms, but include a comprehensive glossary for reference
- Reduce bureaucracy during the Request for Quotation stage. Time is limited and bureaucracy can consume valuable time better spent on solution development and price optimisation
- Allow sufficient time for suppliers to submit their offers. Lack of time typically results in defensive posturing by suppliers, avoidance of innovation, and conservative pricing
- Clearly define the required level of safety/reliability for the equipment (including safety factors and monitoring systems) to facilitate comparable offers
- Anticipate strategy and information to compare offers.

References

[1] Directive 2014/24/EU of the European Parliament and of the Council of 26 February 2014, Annex VI point 1(a).

[2] Global Infrastructure Hub, Reference Guide Output Specifications for Quality Infrastructure Practical guidance with a focus on PPPs and other long-term contracts September 2019.

[3] ICE 'Bridge Deck Erection Equipment: A best practice guide' (Edited by Steve Harridge and Max Meyer).

5.2 In-situ Construction Equipment

Pedro Borges
Portugal

This section offers a concise introduction to the most commonly used deck erection equipment, outlining basic options and concepts. For a more comprehensive understanding of each equipment, see [1–3].

5.2.1 Falsework

Falsework is traditionally used for cast-in-situ concrete deck erection (see Chapter 2.4.2.1). This non-automated equipment consists of a foundation/footing and a supporting steel structure, which can include elements like loose steel scaffold tubes, fittings, adjustable telescopic props, and fork heads, all used to support the deck formwork. The greatest advantage of falsework is its adaptability, stemming from its construction with standard modular components that can be configured in various ways. With high rates of reuse, falsework materials are very cost-effective within their technical application range and have been used in the construction of multi-span bridge decks, arch bridge decks, and single-span bridge decks, among others. However, this equipment is labour-intensive due to the large number of parts that need to be disassembled and reassembled in each erection cycle.

Falsework can be divided into two main categories: Spread Shoring and Portal Shoring.

Fig. 5.2 Spread Shoring example (photo: Construgomes)

Spread Shoring (Fig. 5.2) is heavily reliant on the ground characteristics beneath the bridge, as it requires continuous support for the falsework along the span. It is generally used for bridges with a height of less than 20 metres.

Portal Shoring (Fig. 5.3) reduces the dependency of falsework on ground conditions, as it comprises a portal structure (girders) designed to transfer loads longitudinally. The longitudinal girders in this system are supported by temporary towers that rest on temporary foundations.

Falsework is also commonly employed in combination with other construction methods for erecting bridges with non-standard spans that are not feasible to construct using the main equipment deployed for standard spans.

Fig. 5.3 Portal Shoring example (photo: Jiri Strasky)

Fig. 5.4 MSS example

5.2.2 Movable Scaffolding Systems

A Movable Scaffolding System (MSS, Fig. 5.4) is a type of equipment used in the cast-in-situ, span-by-span concrete deck erection method (see Chapter 2.4.2.2). This method involves assembling all the reinforcement for a span length, casting the concrete for that span, and then post-tensioning it to create a completely self-supported span.

The Movable Scaffolding System (MSS) has evolved naturally from traditional falsework to easily reuse formwork in consecutive spans. This evolution minimises the need for disassembly operations in each deck erection cycle. It also eliminates the risk of relying on direct ground capacity for provisional footings as the MSS is typically supported directly on the previously erected bridge structure. Generally, the MSS is designed to perform the following main functions:

- Support the formwork, the reinforcement steel, and the fresh concrete of the entire span length without intermediate supports. The two most common positions for MSS supports are 1) at the back on the previously erected deck and at the front on a pier, or 2) at both the back and the front on a pier
- Launch the formwork to the next span to be erected using its own mechanical systems, which open and close the formwork to prevent collisions with the piers
- Provide suitable access platforms to all working areas
- Assist the crew in deck erection tasks with internal auxiliary equipment like hoists and gantry cranes.

Most common MSS units are self-propelled, meaning they have their own means of promoting longitudinal movement, including a propulsion system. These units require external lifting and transportation means to move their supports from span to span during each erection cycle. However, state-of-the-art MSS units are self-launching, meaning they are self-propelled systems and, additionally, autonomously advance to the next position without needing external auxiliary equipment for launching, moving their own supports with their internal systems.

Fig. 5.5 Example of Brackets (supports) launching (Credits / Courtesy of NRS AS)

Fig. 5.6 Overhead MSS Main Components example (photo: BERD)

Fig. 5.7 Overhead MSS – front view example (photo: BERD)

Fig. 5.8 Underslung MSS example with Organic Prestressing System (OPS) (photo: BERD)

The main structure of the MSS, which can consist of either a single or multiple longitudinal girders, may be designed to be positioned either above the bridge deck (overhead type) or below the bridge deck (underslung type), see Figs. 5.5 to 5.12. Hybrid solutions, although less common, are also found in some applications.

In each cycle, the MSS erects the deck section between two pier axes (in the case of isostatic spans) or between erection joints. These joints can be positioned at various points along the span, typically located between 1/4 to 1/5 of the span length beyond the front pier. Although less common, multi-span MSS units can also be used to erect two or more spans in a single cycle.

MSS equipment has typically been used in multi-span bridge erection with span lengths ranging from 30 to 70 metres. State-of-the-art MSS are extending the upper span limit and applications with 90 metres of span length have already been used [4].

The duration of the deck erection cycle using an MSS varies greatly, influenced by factors such as the consistency of deck geometry, span length, deck width, type of deck section, bridge design compatibility with the method, crew working shifts, crew experience, degree of equipment automation, extent of rebar preassembly, rate of gain of concrete strength and weather conditions. For reference, at cruising speed, typical working cycle durations range from one week per span (for spans

Fig. 5.9 Conventional Underslung MSS example

Fig. 5.10 Double Span MSS example (photo: BERD)

Fig. 5.11 Overhead MSS – U shape rebar modules (photo: Kappa Kalip Insaat Taahhut A.S)

Fig. 5.12 Underslung MSS – Empty Segments: full section rebar modules with part of inner formwork (photo: BERD)

up to 45 metres) to three weeks per span (for longer spans).

The state-of-the-art MSS are conceived and designed for the implementation of industrialised erection processes, introducing the concept of the MSS as a 'factory'. One of the major advances through the industrialisation of the cast-in-situ process is to design the MSS to be able to receive, handle and assemble the preassembled rebar modules over the formwork.

Furthermore, state-of-the-art Movable Scaffolding Systems (MSS) are equipped with monitoring systems or even active control systems, such as the Organic Prestressing System (OPS) [5]. This innovative system allows the active control of the main structure's deflection during concrete pouring operations (Fig. 5.13), thus extending the upper span range for MSS applications. It has already been successfully applied in projects with spans of up to 90 metres.

5.2.3 Form Travellers

A Form Traveller (FT) is a self-launching equipment that works as a cantilever, supported on the previously erected deck (see Section 2.4.2.3). It is typically used for the cast-in-situ construction of box girders or cable-stayed bridge decks erected as cantilevers. There are two variants:

- Balanced cantilever (Fig. 5.14): erecting the bridge deck by cantilevering segments, starting on both sides of a deck support with a pair of progressive cantilevers. The goal is to keep the loads as balanced as possible between the two sides/cantilevers. It uses a pair of FT units, one at the end of each cantilever tip and, typically, one of the FTs will close the deck between spans.

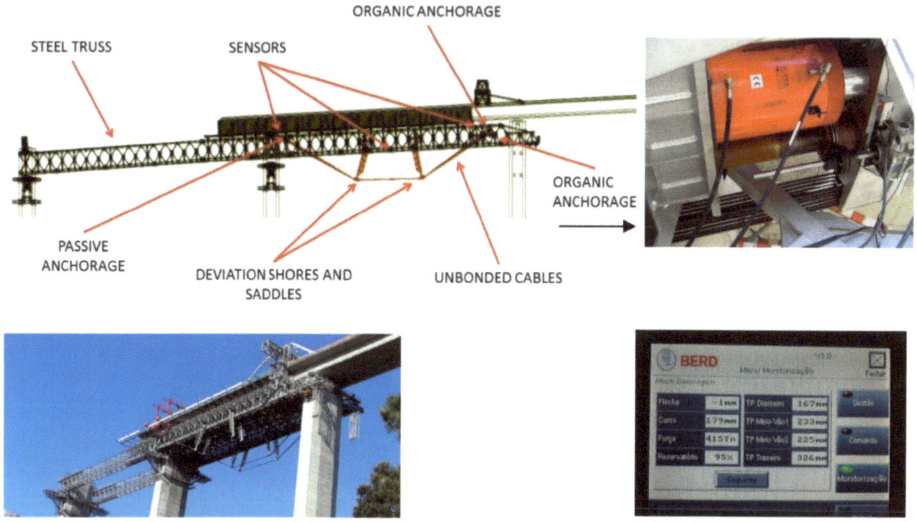

Fig. 5.13 Main components of the OPS technology for underslung MSS [6]

- Progressive cantilever (Fig. 5.15): erecting a bridge deck by cantilevering segments, starting at one end of the deck and progressing towards the other end.

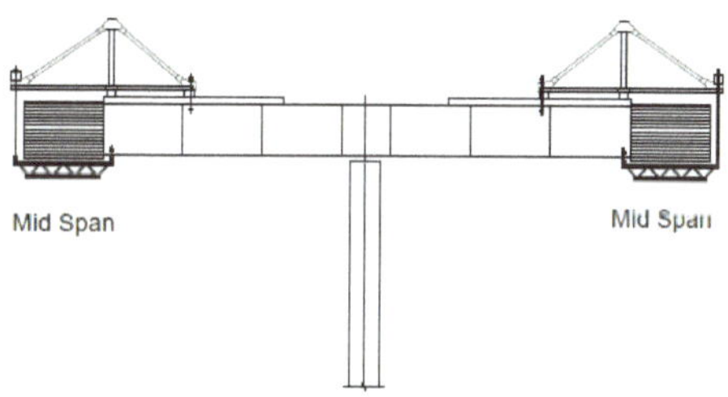

Fig. 5.14 Balanced cantilever

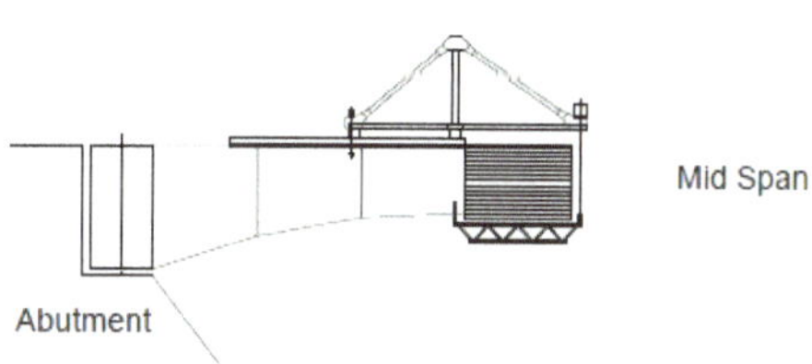

Fig. 5.15 Progressive cantilever

The Form Traveller (FT) is designed to perform the following main functions:

- Support the formwork, reinforcement steel, and fresh concrete.
- Launch the formwork to the position for the next segment's erection.
- Provide suitable access platforms to all working areas including the prestressing platform
- Assist the crew with deck erection tasks (being equipped with auxiliary systems).
- Adjust the formwork alignment to provide the necessary camber

The FT can be conceived to be positioned above (overhead type, Fig. 5.16) or below the bridge deck (underslung type, Fig. 5.17).

The FT is mainly composed of a steel structure, electro-mechanical systems (for levelling, fixing, and launching), the internal and external formwork, and access platforms to the working areas. The FT can have internal hoists to support rebar and deck prestressing tasks. The formwork of the FT is a key component in cases where the deck shape changes

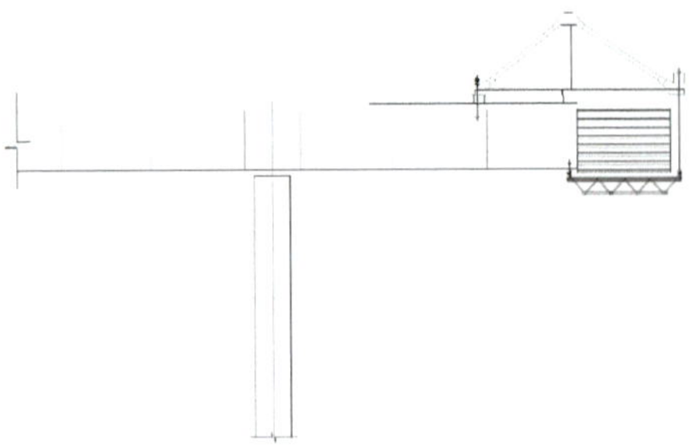

Fig. 5.16 Scheme of an Overhead Form Traveller

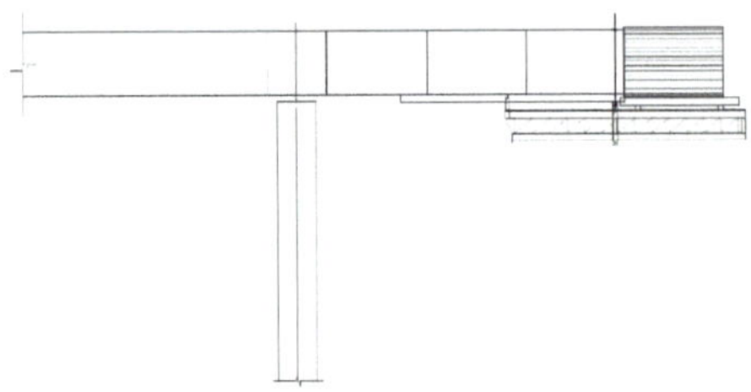

Fig. 5.17 Scheme of an Underslung Form Traveller

along the span (e.g., the deck height usually varies).

The FT is assembled on the deck segment zero (hammerhead) that has been previously erected by other methods. Due to the minimised longitudinal length of the hammerhead (to optimise the cost of its erection), the FT assembly and the construction of the first segment are a special operation. In balanced cantilevers, normally, the first pair of segments is erected with the pair of FT attached together.

The typical FT with two main longitudinal frames permits casting, on average, one segment per week up to 5 metres in length with the segment weighing between 75 to 150 tonnes. In multi-cell box girders, the capacity of the FT can be increased by adding main longitudinal frames, normally aligned with the deck webs. Nevertheless, in more demanding projects, tailor-made FTs are often developed, capable of casting segments up to 12m in length with a segment weight of around 600 tonnes (Fig. 5.18). Some of these high-capacity FTs, applied in cable-stayed bridges (Fig. 5.19), use the deck stay cable to support the cantilever tip of the FT, significantly improving the FT behaviour and reducing the negative moment on the deck cantilever, thus optimising the deck material quantities.

5.2.4 Lessons Learnt and Recommendations

There is a useful statement to reflect on lessons learnt: 'Ignorance is bold'. Keeping this in mind may help to prevent risks in bridge construction. The deck erection tasks should be performed (or supported) by experienced, knowledgeable teams.

- Contractor
- Bridge designer
- Operation team
- Equipment supplier
- Consultants (methods, budget, risk, all to complete the required knowledge).

The engagement of these teams in the project should occur at an early stage, with contributions from all parties, to achieve a compatible, adapted, and optimised solution.

Contractor

Overall savings are the goal. It is important to evaluate whether it is more economical to use more expensive tailor-made equipment (but designed from scratch to be tailored to the specific job, more industrialised, with higher production ratios, self-launching capabilities, lower supplier risks, etc.) than to use standard, existing,

Fig. 5.18 Overhead FT – 10m length pouring (photo: Construgomes)

and consequently cheaper equipment. In some cases, the standard and cheaper equipment is adequate, but significant savings may arise from using tailor-made equipment in special cases where there are large numbers of repetitive spans.

Bridge designer

Nowadays, the powerful computational resources may lead to mistakes, therefore classic design knowledge remains the most important. The bridge design must be fully adapted to the specific construction method chosen. It should consider, from an early stage of the project, the specific characteristics of the equipment that will be used, namely how the equipment performs with deck geometry changes, the interaction with bridge components, and the kinematics. A joint decision, involving all teams, should be made to balance the level of deck material optimisation versus construction simplicity. For example, a lower quantity of prestressing may lead to blisters on the deck, which are complex to remedy.

Fig. 5.19 Underslung FT – deck 34m wide (photo: Construgomes)

Fig. 5.20 Example of a limited surrounding (photo: BERD)

Operation team

Even with a team composed mostly of experienced operators, proper training focusing on the equipment's specific characteristics will reduce the probability of accidents. The implementation of an operation procedure is of utmost importance. To reduce accident risks, the procedure should include the collection and registration of critical data (forces, displacements, among others) to be compared with expected data. Since the operation is dynamic, to increase efficiency, these procedures can serve as a guide for the tasks to be performed and, simultaneously, as a checklist. The checklists should focus on the essentials; too long lists may lead operators to neglect checking.

It is important to maintain consistency in the operation team when using construction equipment with multiple construction cycles (e.g., span-by-span). Frequently changing operators will decrease productivity and quality and increase the risk of accidents. Proper training must be provided for each new operator.

When working in shifts, the transfer of responsibility between shifts should occur at a very defined phase of the tasks. The handover should take place when the shift has completed a full task phase (i.e., not during an intermediate phase of launching). Implementing task checklists at the end and start of shifts will reduce the probability of neglecting or forgetting tasks due to shift changes.

Equipment supplier

The equipment supplier should perform a risk analysis (see Section 4.1) throughout the equipment supply process to identify the involved risks and implement measures to mitigate them. This process should encompass all applicable phases, including the offer, design, manufacture, transportation, testing, assembly, operation, relocations, and final disassembly.

A detailed kinematic study of the equipment, illustrating the full cycle for typical and special deck erections, allows for:

- The identification of the equipment's structural configurations and corresponding loads
- The definition of the equipment's appropriate geometry.

The study of the evolution of the bridge and surrounding site geometry is of utmost importance (Fig. 5.20). For example:

- A study of the deck geometry changes along the bridge is needed to define the necessary formwork shape adjustment mechanisms
- A study of the specific site's available space (in all directions) for the equipment launching to the next position, allowing for the definition of an appropriate equipment geometry, reducing the possibility of clashes with the ground, piers, erected deck, electrical cables, and other potential obstacles.
- Moving loads (e.g., hoist winch trolley, hoist portal crane) require special attention. Detailed kinematic studies along their entire path with the load must be performed. All possibilities of clashes must be identified and mitigated. The defined moving speeds should be set to allow sufficient time to reach maximum speed and to stop safely
- The requirements of the contractor and specialised crew must be valued, as they represent their preferred 'way of doing.' Whenever possible, the equipment supplier should incorporate these requirements, after validating them with a proper risk analysis that considers all dimensions (e.g., safety, kinematics, cost, supply schedule, operation, and assembly).

References

[1] Harridge S. & Meyer M., Bridge Deck Erection Equipment: A best practice guide, ICE Publishing.

[2] *fib* (Fédération Internationale du Béton), Bulletin No. 48, Formwork and falsework for heavy construction, January 2009, (ISBN 978-2-88394-088-8,).

[3] Rosignoli M., Bridge Construction Equipment, 2013, ICE Publishing.

[4] Pacheco P., Coelho H., Borges P., Resende A & Carvalho D. (2021) New Frontiers in Multi-span Prestressed Concrete Deck Construction: A Case Study, Structural Engineering International, 31:1, 106–117, DOI 10.1080/10168664.2020.1738906

[5] Pacheco P., Guerra A., Borges P. and Coelho H. A scaffolding system strengthened with organic prestressing – the first of a new generation of structures, in IABSE, 2007; Vol. 17, Number 4, November 2007: 314–321(8).

[6] Pacheco P., Multi-span Large Bridge Decks – the organic prestressing impact, ISBN 978-1-138-02757-2)

https://doi.org/10.2749/sp001.ch05.3

5.3 Precast Construction Equipment

Pedro Borges

Portugal

This section provides a brief introduction to the most common precast bridge construction equipment, outlining the basic options and concepts. For more in-depth knowledge, please consult [1–3].

5.3.1 Beams Launcher – Beams Erection

The Beams Launcher (BL, Fig. 5.21) is an equipment commonly used in the erection of precast concrete simply supported beam decks (see Section 2.4.2.6). Nevertheless, it can be applied in other contexts as well, for example:

- The erection of precast beams with a construction joint near one-fifth of the span length.
- The erection of precast beams for intermediate sections of large spans with significant longitudinal deck 'hammers' near piers.
- The erection of deck steel girders.

BLs are typically designed as overhead units, and they are equipped to perform the reception, movement, and assembly operations of deck beams with their own BL-powered mechanisms. BLs are typically equipped with:

- One main structure: either two parallel girders or just one girder
- Two lifting devices
- Support frames that are assembled near pier sections, which allow for BL's vertical support and transverse movement.

Fig. 5.21 Beams Launcher example

The beam feed/reception position is often one of the major deciding factors in the design of BL equipment, conditioning the general layout and kinematics.

Considering the erection of span N, four major types of equipment can be defined (Fig. 5.22 to 5.25):

Type 1: Reception of the beams several spans back from span N or from behind the abutment. This BL is launched, transporting the deck beam, and the casting-in-place of the slab is independent of the BL cycle.

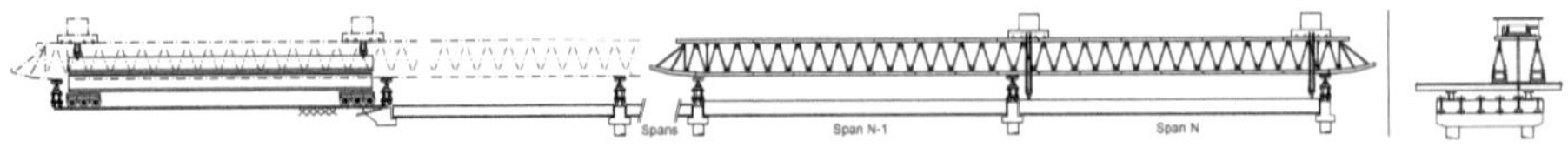

Fig. 5.22 Reception of the beams several spans back of span N or from behind the abutment

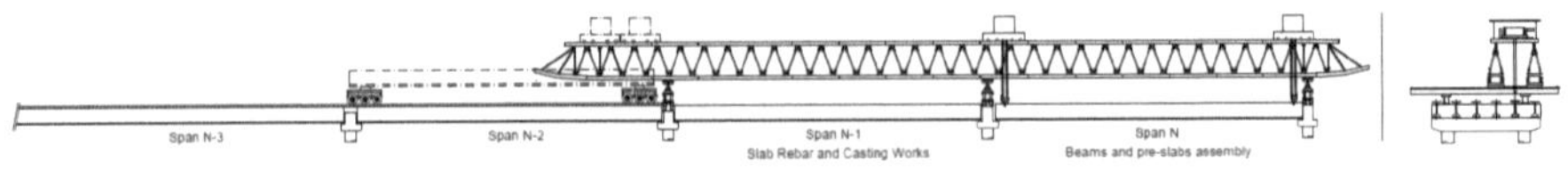

Fig. 5.23 Reception of the beams on the span N-2 and assembly on span N without launching the BL with the beam

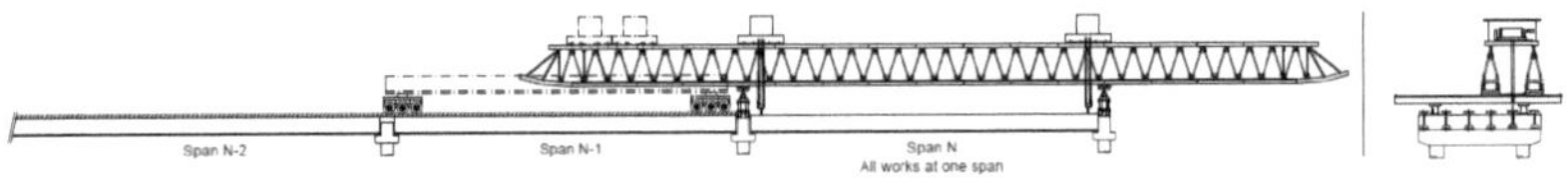

Fig. 5.24 Reception of the deck beams on span N-1

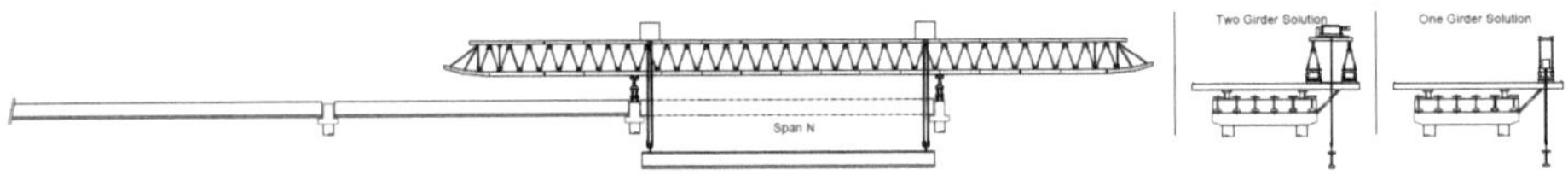

Fig. 5.25 Reception of the beams from below span N under construction

Type 2: Reception of the beams on span N-2 and assembly on span N, without launching the BL with the beam. The casting-in-place of span N-1 is performed during the cycle of beam assembly on span N.

Type 3: Reception of the deck beams on span N-1. The BL cannot operate independently from the slab cast-in-place on span N, due to the lack of a path for beam feeding.

Type 4: Reception of the beams from below the span N that is under construction, most commonly laterally to the deck shadow. This type of BL may also be referred to as a Beam Shifter.

Hybrid combinations of these types can be found to overcome specific feeding issues during deck erection. BLs may also be designed to perform secondary deck erection tasks, such as positioning precast pre-slabs.

5.3.2 Launching Gantries – Segmental Erection

Segmental Launching Gantries (LG) are used in deck erection for the span-by-span construction method, the free cantilever method, or even in hybrid combinations of both (see Chapter 2.4.2.6). In the span-by-span method (Fig. 5.26), all the span segments are temporarily supported by the LG, while in the cantilever method (Fig. 5.27), each segment (or pair of segments) is consecutively assembled at the deck cantilever tips.

Span-by-span LG equipment has been used to construct a wide range of bridge deck types, from smaller and lighter metro rail decks to heavy railway decks with spans of up to 60 metres. Free cantilever LGs are frequently applied on longer-span decks starting from 70 metres. With its own powered mechanisms, the LG can perform the reception, movement (translations and rotations), and fine-tuning assembly of deck segments. Typically, LGs are supported on the previously constructed bridge elements and can self-launch along the bridge's length.

There are several distinct LG models, resulting from the specific challenges of each bridge, the contractor's preferences, and the manufacturer's design 'signature' and know-how. While most LGs are overhead (with main structures over the deck, Fig. 5.28), underslung models (with main structures under the deck, Fig. 5.29) have also been used.

Currently, contractors and manufacturers are pushing LGs towards total automation in performing construction

Fig. 5.26 LG for span-by-span erection with OPS allowing to reduce segments pre-suspension tasks (photo: BERD) [4]

Fig. 5.27 High-Capacity LG for balance cantilevering erection

Fig. 5.28 LG for span-by-span erection Main Components (photo: BERD)

Fig. 5.29 Example underslung LG (Equipment is designed and built by Deal S.r.l a Company of Rizzani de Eccher Group)

Fig. 5.30 LF Example (Equipment is designed and built by Deal S.r.l a Company of Rizzani de Eccher Group)

cycle tasks, aimed at reducing the need for specialised operators, accelerating the cycle, and, most importantly, improving the safety of operations.

State-of-the-art LGs already incorporate several automated systems (e.g., automated connection between the LG and the segment), sensing systems (e.g., measuring the loads on the support frames during the launching operation), and even associated warnings and alarms.

Even though technological evolution in deck construction is traditionally slow, we may not be far from seeing an LG that is fully operated from a central command and control centre, equipped with monitors and all necessary sensing and automation systems.

5.3.3 Lifting Frames – Segmental Erection

Segmental Lifting Frames (LF) are equipment typically used for:

- Cantilever erection of precast segmental decks (Fig. 5.30)
- Construction of cable-stayed decks (see Section 3.1 for this application).

LFs are typically designed as overhead units and are equipped to perform the reception, movements (translations and rotations), and assembly (fine-tuning) of a deck precast segment using their own powered mechanisms. Normally, LFs are self-launching and operate on top of a previously constructed bridge deck, and are typically composed of:

- A main structure, equipped with a load transference mechanism for the deck
- A lifting device
- A longitudinal locomotion system.

Very diverse LF models have been developed, varying significantly in configuration and kinematics. A major factor influencing the equipment's layout and operations is the segment feed/placement position. The most common models are (Fig. 5.31):

Type 1: Reception below the segment's final position (on the cantilever tip)

Type 2: Reception from the already erected deck behind the LF – the LF does not travel with the segment

Type 3: Reception below any longitudinal position of the already erected deck's cantilever shadow – the LF travels with the segment up to the assembly position on the cantilever tip

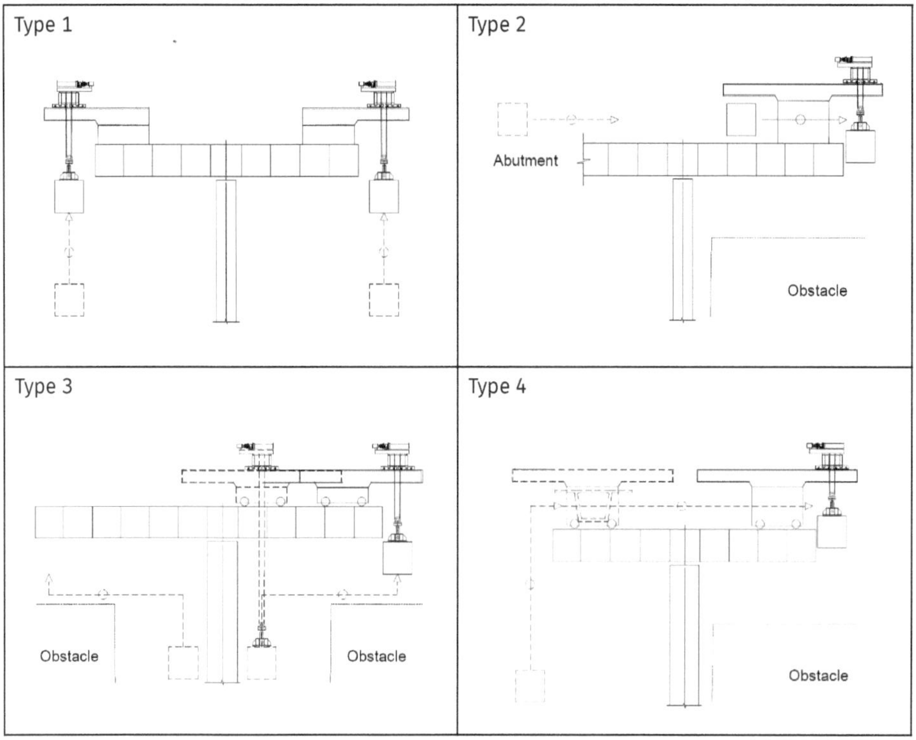

Fig. 5.31 Types of segment feeding in cantilever construction

Type 4: Reception on the opposite cantilever side in balanced cantilevering construction – the LF travels with the segment up to the assembly position on the opposite cantilever tip.

5.3.4 Lessons Learnt and Recommendations

Lessons learnt of Chapter 5.2.4 are applicable.

References

[1] Harridge S. & Meyer M., Bridge Deck Erection Equipment: A best practice guide, ICE Publishing.

[2] *fib* (Fédération Internationale du Béton), Bulletin No. 48, Formwork and falsework for heavy construction, January 2009, (ISBN 978-2-88394-088-8,).

[3] Rosignoli M., Bridge Construction Equipment, 2013, ICE Publishing.

[4] Pacheco P., Resende A., Torres A. & Coelho H., Span by Span Segmental Construction: OPS Impact in the LG50-S Laguna Case, IABSE Symposium, 19–21 September 2018, Nantes, France.

https://doi.org/10.2749/sp001.ch05.4

5.4 Special Heavy Lifting Equipment

Javier Martinez

Spain

This chapter will examine how heavy lifting equipment can support alternative, smart, and innovative construction solutions for major civil structures, particularly bridges. Bridges, by definition, connect two points separated by a geographical obstacle such as a road, valley, river, or bay, and are installed in a challenging environment with various constraints and limitations.

Special heavy lifting equipment is a challenging and complex technology that requires specialised designers, contractors, and suppliers. This section presents a very brief overview of some of the relevant aspects regarding this specific type of technology.

Thanks to the availability of heavy lifting equipment, in many cases, a bridge or its components can be assembled and built in a different location from its final position. This can make it more accessible for cargo segments and workers and allow for ground-level construction without permanent limitations on height, traffic, environment, worker safety, or any other specific constraints.

Traditionally, larger and smarter heavy lifting equipment has been designed for the oil and gas industries, which have much higher budgets and handle much heavier loads than the civil construction industry. In fact, the heaviest loads handled by the oil industry are often ten times heavier than those in the civil construction industry.

It is usually mentioned that the heaviest loads handled by the oil industry are one zero heavier than civil, i.e., ten times, and that parameter has been held for the last forty years. Reference weights handled onshore for the oil industry over the past years are as follows:

+ over the 1980s: 3,000 tonnes/unit
+ over the 1990s: 10,000 tonnes/unit
+ over the 2000s: 20,000 tonnes/unit
+ over 2010/2020: 40/50,000 tonnes/unit

However, in recent years, heavy lifting equipment has become much smarter, more efficient, and more cost-effective, with wireless operational systems and controls making it much more accessible to the civil construction industry.

To take full advantage of these innovative tools, construction engineers must also define alternative construction solutions. They must possess engineering expertise and knowledge of heavy lift handling to optimise the use of these tools in specific cases for this industry.

This chapter will analyse the different typical jobs where heavy lifting equipment and engineering expertise have been used in the civil construction industry.

Fig. 5.32 Launching of two bridges over Beauharnois Canal in Montreal, Canada

5.4.1 Bridge Incremental Launching

The proposed method involves constructing a bridge entirely in one of two launching bays at ground level, with convenient access for transportation of bridge parts via road, the use of small cranes, accessible bolting and welding at ground level, and easy and efficient inspections. This approach can be applied to steel, composite, or concrete bridges of constant depth, and in certain specific cases of variable depth. This construction method is particularly advantageous and recommended for bridges with multiple spans, high vertical clearance, straight or constant bends with a radius of more than approximately 600 metres, slopes up or down, and spans longer than 40 metres but up to approximately 160 metres for steel/composite bridges. Some examples in Fig. 5.32 to 5.34.

5.4.2 Dismantling & Installation of Single Segments by SPMT

The need to install or remove bridge segments is becoming increasingly common, especially in areas of congested traffic in

Fig. 5.33 Launching of Caracas – La Guaira Viaduct, Venezuela

Fig. 5.34 Launching a second bridge over the Orinoco River, Venezuela

Fig. 5.35 Transport and installation of Concordia Bridge in Madrid, Spain.

Fig. 5.36 Replacement of La Paz bridge spans in Madrid, Spain

Fig. 5.37 Installation section of Rego de Lamas Viaduct in Orense, Spain

Fig. 5.38 Installation of Sky bridge at Hong Kong International Airport, China

Fig. 5.39 Installation of Edmonton's light rail elevated guideway

cities or on busy roads. This construction alternative allows for the installation or removal of segments of any length and weight in a short amount of time, typically overnight, minimising traffic disruptions. Self-propelled modular transporters (SPMT) are used to remove old bridge spans and transport them to an area free of traffic, where they are dismantled and replaced with new segments. Ideally, bridge segments with a vertical clearance between 4.5 and 9 metres are suitable for this method, although higher heights can be managed with self-stable jacking systems installed on top of the SPMT. This method is widely used in Europe and the United States, with national design and construction specifications developed to estimate the dynamic effects on the bridge and temporary supporting structures during transport. Ground bearing limitations must be analysed when using this method in downtown areas where there may be underground infrastructure. The typical nominal bearing pressure transmitted by an SPMT is just below 10 t/m^2, but more SPMT lines can be added if less pressure is needed. The ground can be reinforced with steel plates or scaffolding if lower pressure is required. Some examples in Fig. 5.35 to 5.39.

Fig. 5.42 Lifting of Vicaria bridge in Albacete, Spain

5.4.3 Simple Span Lifting

When heavy equipment cannot access underneath a bridge span or when the clearance is very high, and/or lifting is required over water, strand lifting units , as the ones presented in Fig. 5.42 to 5.44, can be installed in two consecutive piers. This allows for simple segment lifting jobs, lifting the corresponding span. In such cases, there are no restrictions on height, weight, or span length, and the possibilities are practically unlimited.

Fig. 5.43 Lifting of three steel walkways at Tous Dam in Valencia, Spain

5.4.4 Cable-Stayed Bridges: Lifting Segments in Cantilever

This construction method can be used for steel and concrete segments of various lengths and weights, with common lengths around 10–12 metres for concrete and 10-20 metres for steel. Capacities are up to 1000 tonnes, and higher capacities can be evaluated. Traditionally, large lifts have been performed by gantries installed on the deck, which lift the segments vertically from below, whether the segment is above land or water. In recent years, bridge owners have imposed new constraints that make it difficult to access the area below the bridge, such as road access restrictions, obstructed water channels, high water currents, or freezing water in winter. As a result, design-

Fig. 5.44 Lifting of the removable span of 1812 Constitution Bridge in Cadiz, Spain

ers have had to increase the lifting capacity of traditional lifting elements and design high-capacity, electronically controlled, 360-degree rotation devices to allow the installation of heavier loads from above the deck. Two examples in Fig. 5.45 and 5.46.

Fig. 5.45 Lifting deck sections of 1812 Constitution Bridge in Cadiz, Spain

5.4.5 Suspension Bridges: Lifting Deck Segments

This bridge design is not yet extensively improved by standard modular heavy lifting equipment. The main suspension cable system is installed using well-established procedures, and the only significant weights to be installed are the segments. The size and weight of these segments are limited not only by the capacity of the equipment used but also by the geometry and longitudinal distance between lifting hangers. Achieving homogeneous equilibrium when lifting two symmetrical segments to properly load the cable catenary can also be a challenge. See example in Fig. 5.47.

Fig. 5.46 Lifting closing deck section of second bridge over the Orinoco River, Venezuela

5.4.6 Installation by Transverse Launching

This system is particularly useful in cases where access to the bridge is difficult and a single or multiple spans need to be replaced quickly. It is a highly regarded solution for rail bridges where trains need to resume operation in a very short period, usually within 12 hours or less. By constructing the new bridge or span parallel to the old one, the lateral shift can be completed within this tight time frame. The bridge is typically slid along a temporary rail system using hydraulic jacks or cable strands. This construction method is also cost-effective and entails less risk. Some examples in Fig. 5.48.

Fig. 5.47 Lifting sections of Osterøy Suspension Bridge in Norway

Fig. 5.48 Transversal skidding of railway bridges, several locations, Spain.

5.4.7 Installation of Single-Span Bridges Over Water

When a single-span bridge with a span length of up to 250/300 metres needs to be installed or removed over water, there are various solutions available depending on the manufacturing location of the new bridge (Fig. 5.49). If the new bridge can be manufactured on a launching bay at one abutment, it can be assembled there, or it can be fully assembled away from the site and loaded onto a barge for transportation. The vertical clearance of the bridge at the permanent position is also an important factor to consider. If the clearance is up to a maximum of 30 to 40 metres, a new high-powered self-climbing jacking system can be installed on a barge deck. If the clearance is higher, strand lifting units can be installed on a cantilever frame on top of the side spans.

5.4.8 Building Arches or Piers Vertically and Tilting Them

When constructing a single-span bridge with an arch support, whether made of steel or concrete, a major challenge is the construction of the lower arch that supports the entire deck. Once the arch is complete, installing the deck is not a problem.

The traditional method of building the arch involves constructing it in successive cantilevers suspended from two strong temporary anchor masts on each side. This method is widely used, but can be slow when constructing a concrete arch and complex to maintain proper geometry when building a steel arch.

Another option is to construct the arch in two halves, one on each side, in a verti-

Fig. 5.49 Single-span bridge installation, different typologies and techniques

Fig. 5.50 Overview of different arch tilting examples, Spain

Fig. 5.51 Jacking-up of Loenerslootse Bridge in Amsterdam, The Netherlands

Fig. 5.52 Jacking-up Danube Bridge in Linz, Austria

cal position, and then tilt them into place by swivelling the lower support, pushing them into place, and holding them until they achieve the correct shape (Fig. 5.50).

5.4.9 Jacking up Segments or Full Decks

One alternative that heavy lifting equipment allows is to construct a bridge, segment, or full span just below its own shade and then jack it up to its final position (Fig. 5.51 and 5.52). Traditionally, available jacking equipment in the market only allowed for safe jacking-up of segments or decks up to a height of 2.5 to 3 metres, which is too short as the minimum vertical clearance of a bridge is typically around 5 metres. Therefore, in the majority of the cases, the moving up operations had to be performed by lifting. However, in the past five years, several self-stable jacking systems of various capacities have been developed, making it easy to lift structures to greater heights.

5.4.10 Concrete Girder Launchers

There is a wide range of different tools and equipment to use, which is detailed in other chapters.

5.4.11 Cranes

This chapter only discusses cranes with a capacity of over 1000 tonnes. Over the last 15 years, the crane design sector has made impressive developments, driven by the needs and requirements of the offshore, oil and gas, and nuclear industries for high-capacity cranes. At the end of the 20th century, the largest cranes had a maximum arm capacity of 12,600 tonnes × metres and a maximum lifting capacity of around 1000 tonnes at short radius. Today, cranes are available in the market with up to 350,000 tonnes × metres of arm capacity and 5,000 tonnes lifting capacity at short radius, and new developments are underway. It can be foreseen that cranes with even higher capacities will be available in the market in less than three years.

The oil and gas industry requires such cranes to work within short radii, lifting modules up to 5,000 tonnes within a 60-meter radius, mainly for onshore installation of floating production storage and offloading unit (FSPO) modules. While the civil construction industry typically does not require such high loads, this new tool is being used extensively for much lower weights but larger radii, for example, lifting 200 tonnes over a radius of 180 metres (Fig. 5.53).

In addition, these new cranes have been designed to be containerised and reach any location worldwide. Transport and mobilisation or demobilisation times are shorter than those of standard large crawlers.

Fig. 5.53 Overview of different types and configurations of ringer cranes

5.4.12 Miscellaneous

'Fishing' a bridge

Many heavy lifting challenges in civil engineering can now be resolved by using a smart combination of various types of equipment, systems, and methodologies mentioned in this book. This section will showcase some civil projects that have been completed in recent years.

One example is 'fishing' a bridge: when installing a single-span bridge over a river, if there is no access to a barge or guaranteed minimum draft, the front section of the bridge can be 'fished' from the opposite abutment (Fig. 5.54).

Erecting an arch

For this installation of an arch bridge over a dam with no guaranteed water level and a long span, one solution is to build the

Fig. 5.54 Launching of the Pavilion Bridge in Zaragoza, Spain

Fig. 5.55 Skidding and tilting arch section of Alconetar Viaduct over Alcantara Dam, Spain

Fig. 5.56 Skidding of an entire palace, Mexico City

arch in four segments onshore, with two segments on each side. The segments can then be slid over the deck and subsequently tilted into place and connected (Fig. 5.55).

Shifting a building

A downtown palace situated in a major, crowded city had to be temporarily removed from its permanent position to prepare for important public ground infrastructure. To accomplish this, the palace was hydraulically skidded away, providing clear access to the ground underneath. Afterwards, the palace was replaced back into its original position (Fig. 5.56).

5.4.13 Lessons Learnt and Recommendations

The heavy-lifting sector is constantly evolving, and there are countless lessons to be learnt from every job. Just a few examples are indicated below:

- It is key to ensure an early project involvement, even at the conceptual phase, of all necessary parties in the decision-making process, including the designer, construction company, and the heavy lift specialist.
- There must be a proper choice and definition of the project solution and the required set of temporary works, which can have a minimal cost impact if the decisions are made in the early stages of the project, even before the preliminary foundation works have started.
- The total weight to be lifted is not always the most critical factor when choosing heavy lifting equipment. While heavy lift tools are often designed for the oil and gas industry, where weights are much heavier than in the civil construction sector, other factors such as overall dimensions, geometry, centre of gravity position, height, distance to be lifted, required temporary works, and integration into the larger project must also be considered to minimise the impact.

References

[1] General Archives of Lastra Iberica, S.A.
[2] General Archives of ALE Heavylift Iberica, S.A.
[3] General Archives of Mammoet

5.5 Monitoring During Construction

Filipe Magalhães[1] | João Santos[1] | Jon Ruiz de Azúa[2]

[1] Portugal, [2] Spain

5.5.1 Introduction

This section addresses monitoring during the construction of bridges. It presents the available tools, describes what can be achieved with different monitoring components and solutions, provides examples of monitoring strategies for alternative construction methods, and ends with some expected future trends and critical lessons.

5.5.2 Monitoring Systems Technology

5.5.2.1 Monitoring Systems Main Components

Traditional monitoring systems are composed of a set of sensors connected by cables to a central digitizer, which converts analogue electrical signals into time discrete readings, and a central computer that organises the measurements in text files or databases, which might have local data processing abilities, remote internet connection for data download, and data viewing tools.

With a view to cable length reduction and mitigation of electrical noise (associated with long sensor cables), the digitalisation can be distributed, grouping subsets of sensors in several digitizers connected by a single cable (that can assure power, time synchronisation, and digital data transmission) to a central station. However, the continuously changing conditions associated with bridge construction, make the use of cables quite troublesome. Therefore, this is an application where the use of wireless solutions is clearly advantageous. In this scenario, the monitoring system is based on nodes, which normally contain one or several sensors, a digitizer, possibly local memory for data backup, and hardware for data transition to a central station. In cases where cables need to be completely avoided, batteries should also be included. With the current technology, if sampling rates of some minutes (suitable for static measurements) are combined with a smart use of data transmission, it is possible to have some months of autonomy. Ideally, the central station is supported by a power supply and connected with a modem to send the collected data to a server, where the data processing is performed, the results are stored and made available to the users through a continuously updated web page.

In some case studies, the monitoring system adopted during construction is converted, with adjustments, to a long-term monitoring system for the built structure. In this case, solutions based on cables might be more convenient, as

Fig. 5.57
Sensors for the characterisation of external loads and conditions: (a) force measurement using a pre-calibrated instrumented bearing device, (b) ultrasonic anemometers used in a movable scaffolding, (c) force measurement in a hanger, (d) geophone vibration sensor, (e) embedding thermometers, (f) creep and shrinkage specimens, (g) sonar scour sensor and (h) radar water level sensor [1].

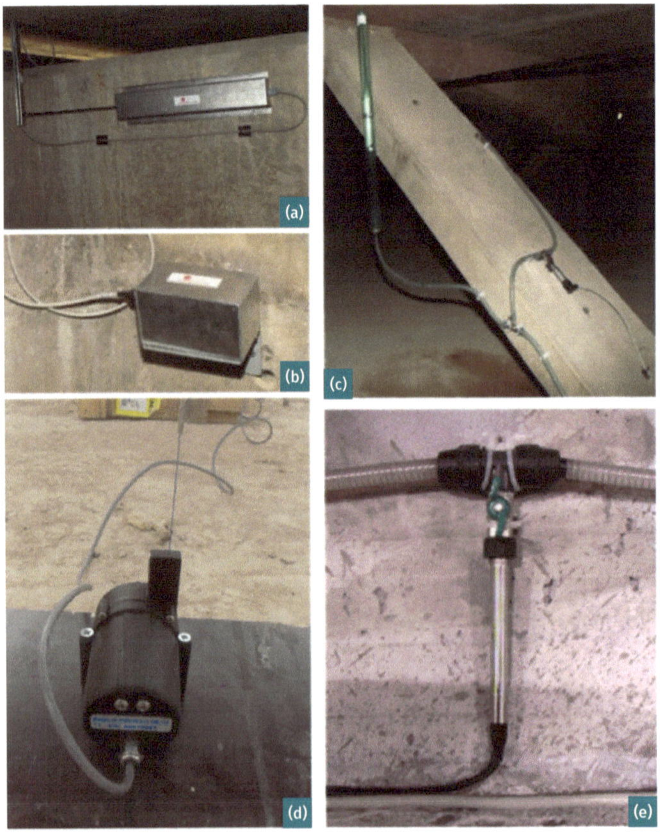

Fig. 5.58
Measuring displacements: (a) magnetostrictive transducer for relative displacement measuring [2], (b) servo tilt-meter for static rotation measurement [2], (c) hydrostatic levelling system for relative vertical displacement measurement [2], (d) precision potentiometer for relative displacement measuring, (e) absolute pressure cell.

the necessary maintenance is low when compared to wireless solutions that need battery replacement.

5.5.2.2 Sensors

The currently available sensors allow measuring all the relevant quantities related to the loads and subsequent structural response with good accuracy.

During construction, the most relevant loads are associated with self-weight, temperature variation, and wind. Wind characterisation is common, or even mandatory, in all construction projects where high cranes or piers are present. It might also be relevant to monitor accidental loads, motivated by impacts, and ground vibrations caused by rock blasting, heavy construction activities, or earthquakes. Fig. 5.57 presents some examples of sensors for load characterisation.

In the characterisation of the structure's response to the applied loads, the following quantities can be measured: displacements, internal forces (mainly axial forces and bending moments), strains, and accelerations.

During bridge construction one of the main concerns is geometric control, therefore implying the paramount importance of the measurement of displacements. This can be done with topography tools, such as total stations, either with hydrostatic principles or directly measuring relative displacements with linear variable differential transformer (LVDT) transducers. Fig. 5.58 shows some examples.

The measurement of internal forces (Fig. 5.59) in deck and pier cross-sections is commonly performed using 4 strain gages (these might be electrical, fibre optic, or vibration chord) in relevant cross-sections, allowing the determination of the axial loads from the mean values and the moments in each principal direction from strain differences. Still, variables involving less localized measurements are normally easier to interpret and less prone to installation inaccuracies. For instance, with two clinometers installed in two close sections of a bridge deck, it is possible to calculate the curvature and obtain bending moments; with the estimation of the main natural frequencies of vibration of a cable, by means of an accelerometer, it is possible to estimate the installed axial force (using the vibration chord theory).

5.5.2.3 Monitoring Goals

A monitoring system may include several components, which should be combined in a smart way, depending on the specific characteristics of each application.

In the previous section, the tools for load characterisation have been presented. Regarding structural response, the focus could be on the response to static loads or to dynamic loads.

During construction, the response to static loads is usually more relevant. In this case, sampling rates of several minutes are appropriate to characterise the structure's response. The analysis of fast variations (less than a few hours), associated with time-constrained and well-defined operations is quite straightforward. Data processing includes only the conversion to engineering units and possible data cleaning (for instance by means of moving averaging). For the analysis of periods from several hours to many years the data processing should include tools to separate the contribution of the loads, temperature, and rheological effects (creep and shrinkage in concrete).

Fig. 5.59
Measuring internal forces: (a) and (b) mono-strand stay-cable load cell [3], (c) embedding strain gage, (d) and (e) accelerometers capturing cable/hanger vibrations and (f) tilt-meter.

The analysis of the structure's dynamic response implies the use of high sampling rates (several measurements per second). It might be useful to evaluate vibration levels due to activities in the vicinity that may for instance influence the concrete curing or to evaluate the time variation of the natural frequencies of relevant structural elements (an indirect mean of evaluating the time evolution of their stiffness). For the last goal, algorithms for automated operational modal analysis should be used [5].

A very relevant topic during bridge construction is the control of the material properties, especially in the case of prestressed concrete elements, where the time evolution of the concrete elasticity modulus and of the deformations due to creep and shrinkage can have an important impact on the construction pace and geometric control. The time evolution of the modulus of elasticity is usually controlled with compression test in samples; shrinkage and creep might be experimentally determined, monitoring the time evolution of the deformation of samples placed inside the deck (a free specimen in the case of shrinkage and a compressed specimen for the case of creep evaluation), see Fig. 5.57f.

5.5.3 Construction Methods Versus Monitoring Layouts

This section presents some relevant case studies for the purpose of illustrating the monitoring layouts that are most appropriate for each construction method.

5.5.3.1 Balanced Cantilever

In cantilever construction (described in Section 2.4.3.4) rigid body motion and overstress in the upper area of the deck cross-sections should be controlled. Geo-

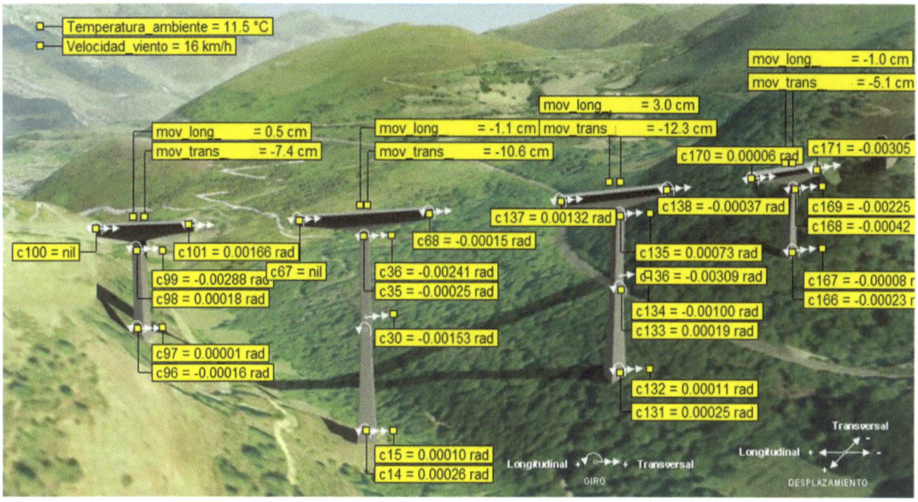

Fig. 5.60 Online monitoring of the displacements (values in cm) and rotations (values in rad) of Montabliz Bridge (courtesy of Kinesia Data)

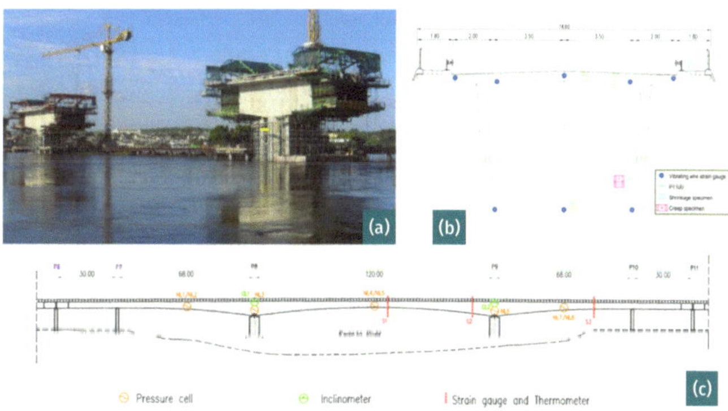

Fig. 5.61 17 de Setembro Bridge, in Cabala, Angola [1]: (a) overview of the balanced cantilever method, (b) embedded sensor locations in the girder's concrete section, (c) overall instrumentation view.

detic measurements are usually put in place and taken as frequently as possible, to ensure that no rigid body motion is observed. However, tiltmeters and hydrostatic levelling systems composed of a fixed cell over the piers and a movable cell at the front end of the cantilevers also consist of a simple, effective, and cost-efficient approach (see Fig. 5.60). Overall stress can be controlled by installing strain gages at predefined sections, as shown in Fig. 5.61. However, care must be taken for all effects associated with concrete hardening and rheology, as the errors of not modelling these effects properly may be of the same order of magnitude as those that must be observed.

Fig. 5.60 and Fig. 5.61 present two examples of monitoring systems of balanced cantilever bridges.

In short, the most relevant monitoring components for this construction method are:

- clinometers at the foundation, top of piers, and 1/4 of the span
- temperature gauges installed inside the concrete (at least 4 per instru-

mented cross-section) to characterise the thermal actions in the deck and piers (only relevant in the case of high piers)
- measuring stresses in the reinforcement bars or concrete is not as useful for controlling the construction stages as measuring the rotations, but these are commonly considered
- if an adequate number of clinometers is installed, the approximate vertical deflection of the cantilever can be monitored in real-time.

5.5.3.2 Incremental Launching

Incremental launching construction processes (described in Section 2.4.3.3) frequently generate the largest stresses that the launched beam or girder will experience throughout its life cycle. For this reason, care must be taken to measure these quantities, which can increase over expected ranges, if local buckling and excessive deformations are observed. Therefore, the monitoring should include the measurement of longitudinal stresses and rotations at predefined sections as well as forces at the supports to ensure overall displacement control of the launching. Rotations should also be measured at the front end of the launched beam [4] and at the top of the piers. These quantities must be previously estimated numerically to allow for obtaining real-time evaluation. In steel decks, transverse stresses must also be measured at predefined critical sections to ensure the absence of local buckling and for early identification of distortions in the launched beam. A comparison between previously calculated numerical values of rotation and stresses observed in a case study is shown in Fig. 5.62. Fig. 5.63 displays an interface for online control of the rotations at the top of the piers (in the transversal and longitudinal directions) during launching.

Operations should also be controlled through the monitoring of the pushing jacks' oil pressure. Validation using geodetic measurements is usually conducted and considered good practice.

5.5.3.3 Cable-stayed Bridges

The construction of cable-stayed bridges is discussed in Section 3.1. The monitoring layout will depend on the bridge size and particularities. However, the basic set of sensors includes:

- clinometers on piers, possibly at foundations, and at the top
- temperature gauge on pier/tower cross-sections for assessment of thermal actions
- temperature gauges on deck sections for assessment of thermal actions
- temperature gauges on cables
- force measurements on cables, either using load cells for individual cables or indirect measurements using accelerometers to identify natural frequencies and estimate the force using the vibration chord theory
- strain measurements at pier cross-sections (optional)
- strain measurements at deck cross-sections (optional).

Fig. 5.64 shows an example of a monitoring system used during the construction of a cable-stayed bridge and then converted to a permanent monitoring system of the structure.

5.5.3.4 Bowstring Bridges

The elevation of the arches in this type of bridge imposes stress distributions to these elements that are significantly dif-

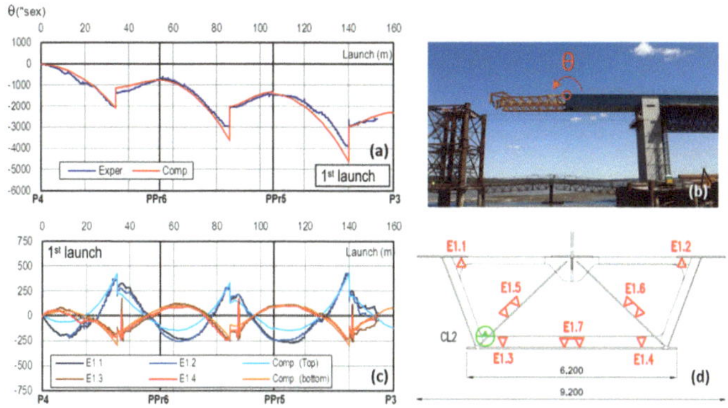

Fig. 5.62 Railway crossing of the River Sado incremental launching, Portugal [4]: (a) and (b) rotations measured at the front end of the launched girder, (c) and (d) embedded sensor evolution and locations in the girder section

Fig. 5.63 Regueirón Bridge, rotation (in rad) at the top of each pier (courtesy of Kinesia Data)

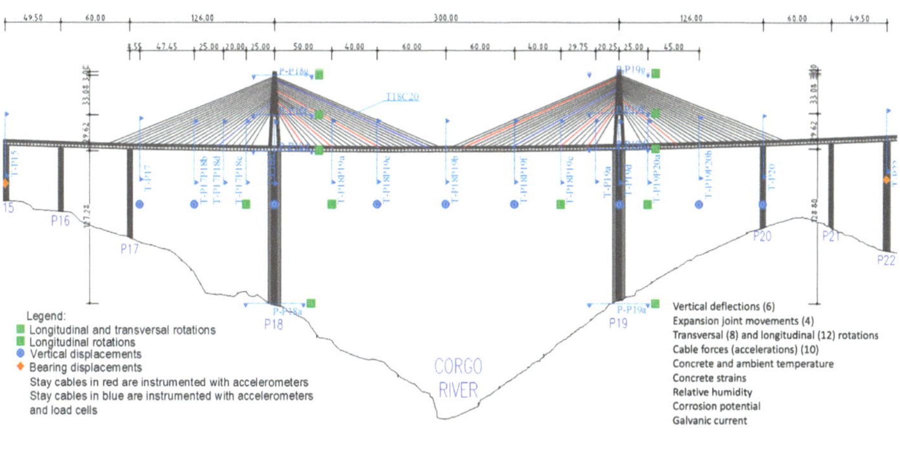

Fig. 5.64 Monitoring layout of Corgo cable-stayed bridge used during construction and operation [6]

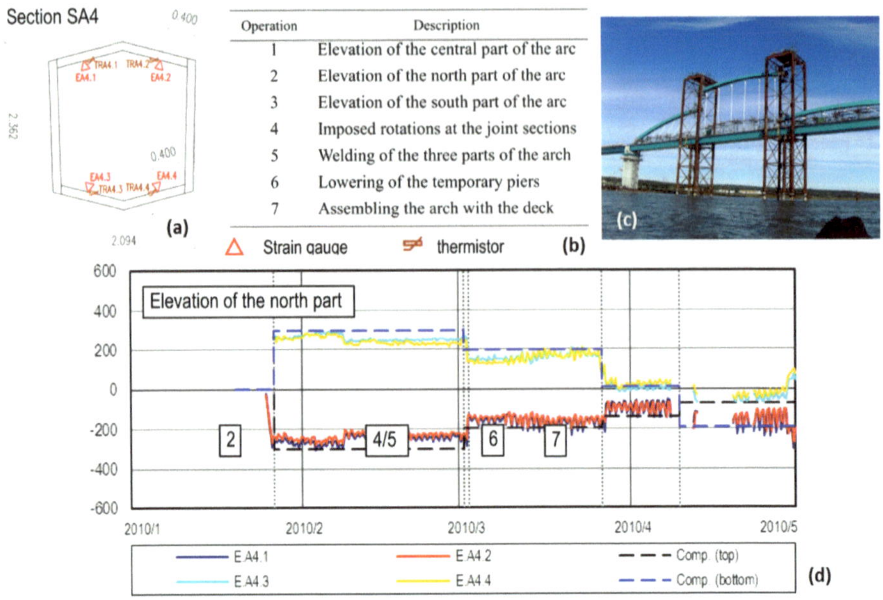

Fig. 5.65 Arch hoisting of the Railway Bridge over the River Sado, Portugal [4]: (a) arch cross-section and instrumentation, (b) hoisting operations, (c) overview, and (d) stress evolution against reference numerical results (comparison).

ferent from those observed during service. Hence, their design and construction plans face the challenge of avoiding excessively large stresses as a consequence of the construction. These challenges motivate designers to consider lower ratios of resistance/actions during arch hoisting given that the process is well controlled. Under these circumstances, monitoring is particularly necessary, not only to ensure real-time safety but also to validate that no local yielding or excessive deformation is observed. As for the incremental launching, stresses and rotations should be measured at certain predefined sections previously studied numerically. Hydrostatic levelling is generally not an option due to the significant height differences observed between sections, and tiltmeters must be chosen carefully since several of the most precise options, such as those based on servo sensors, generally work for certain predefined rotation domains very close to the horizontal plane.

Unlike incremental launching, arch hoisting can vary significantly regarding the nature and order of the operations involved and therefore the monitoring systems must be developed accordingly. Wireless sensors are highly recommended, as was the case for hoisting the central subsets of the Sado Bridge's arches [4], shown in Fig. 5.65. In this figure, the correlation between experimental and numerical measurements can be observed.

5.5.3.5 Arch Bridges

Arch bridges can be built using very different construction methods (as presented in Section 3.3), but a few similarities in monitoring can be noted:

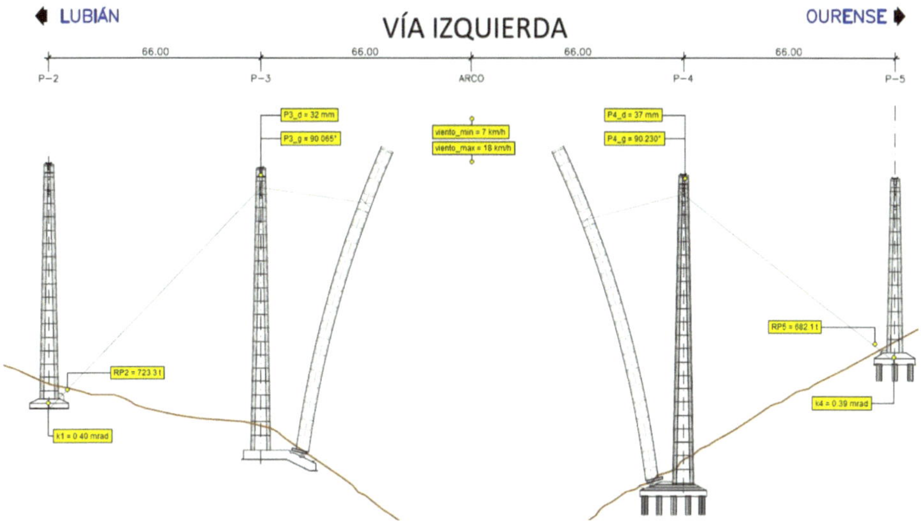

Fig. 5.66 Teixeiras Viaduct online monitoring: cables force in tonne (t), piers base rotation in mrad, piers top displacements (in mm) and rotation (in º) (KSM Structural Monitoring courtesy)

- longitudinal tiltmeters on 1/2 length of arch
- longitudinal tiltmeter on temporary piers supporting the arch
- auxiliary structure from which arch is hanging (cables) instrumented with load cells, strain gauges, or accelerometers for force estimation
- thermal action in some deck and arch cross sections.

Case study 1 – Teixeiras Viaduct

This arch was built vertically and then placed into its final position. The monitoring consisted of (Fig. 5.66):

- tiltmeters on back piers (the ones from which cables hold the arch) that allowed the calculation of their horizontal deflection
- strain gauges on cables to characterise cable force
- accelerometers on cables to obtain frequency and then force
- tiltmeters on foundations where cables are anchored
- steel cables temperature
- pier temperature gradient.

Case study 2 – Infante D. Henrique Bridge

The high complexity of the construction process of this bridge required the design of a very comprehensive monitoring system that provided the characterisation of a number of different parameters, such as reactions in supports and bending moments, axial stresses, rotations, and temperatures in selected sections of the arch, deck, columns, and provisional struts and pillars. A total of 120 sensors were installed (strain gauges, tiltmeters, thermometers and load cells), in the section highlighted shown in Fig. 5.67 and the forces in the temporary cables were regularly measured with accelerometers.

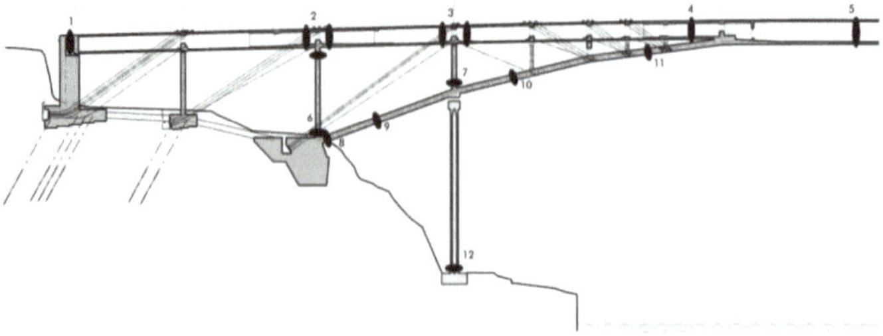

Fig. 5.67 Part of the Infante D. Henrique Bridge instrumented during construction [7]

5.5.3.6 Suspension Bridges

The complexity associated with the building of suspension bridges (as described in Section 3.2), as well as their flexibility and low structural redundancy places this type of structures as those which are more difficult to monitor, not only during service but also, and most particularly, during construction. The stress levels measured during service depend directly on the values observed during construction. Processes such as the cable installation or the hanging of stiffening truss sections are complicated and worth monitoring through the stresses in the newly installed structural elements as well as through the rotations on the anchoring blocks and in all existing deviation cells along the cable's length. However, the definition of the operations for force adjustments throughout the hangers and cable sections and the definition of the final geometry are of particular importance and, therefore, in addition to rotations, displacements must also be controlled, using hydrostatic levelling systems and/or geodetic systems. Overall stress measurement is encouraged and useful not only for validation but also for future life-cycle analysis and maintenance or repair. Given that it is generally a relevant action for these slender structures, wind should also be measured during construction.

5.5.4 Lessons Learnt and Recommendations

The currently available technology has been contributing to making monitoring less expensive, much easier to deploy, and more effective than in the past . The lower costs associated with electronic components and the available wireless protocols support the development of cost-optimised and customised solutions; the expanding availability of high-speed internet makes remote access to large amounts of experimental data a reality; and finally, the present data processing capabilities allow the online conversion of large amounts of data into useful information for the contractor, bridge design team, and bridge owner.

Past experience has demonstrated that the costs related to monitoring equipment have dropped substantially, to the extent that nowadays they are irrelevant in the construction of large infrastructures.

The bridge construction industry should fully endeavour to follow the technological revolution, harnessing it to reduce risk, optimise construction times, and improve knowledge.

Instrumentation should be used not only to monitor and control the structural behaviour of the bridge under construction but also to evaluate the performance of construction equipment and characterise the interaction between equipment and bridge. The monitoring of construction equipment is still not common, but with a smart design of monitoring systems for these machines, it is possible to increase safety (with better control of loads), optimise their design (with the evaluation of design assumptions), improve life-cycle analysis (with the recording of stress time histories), optimise maintenance/repair actions and even decrease construction time (for instance with the indirect characterisation of concrete hardening).

The monitoring strategy should be established at an early phase and agreed upon by all the stakeholders involved in a drive to enhance the synergies between the requirements for the construction, for the performance of commissioning tests, and for observation during service.

References

[1] L. O. Santos, X. Min, J. P. Santos, and M. M. D'Abril, "Structural Health Monitoring of Cabala Bridge in Angola Bridge," in IABSE Conference – Global Thinking in Structural Engineering: Recent Achievements, 2012, no. 774

[2] J. P. Santos, "Smart Structural Health Monitoring Techniques for Novelty Identification in Civil Engineering Structures," PhD Thesis. Instituto Superior Técnico – University of Lisbon, 2014.

[3] V. Marecos, "Controlo da Segurança de Ponte em Tempo-Real," Technical University of Lisbon, Instituto Superior Técnico, 2007.

[4] L. O. Santos, M. Xu, and J. P. Santos, "SHM of the new bridge over the River Sado in Portugal during construction," in Symposium IABSE, Long span bridges and roofs: development, design and implementation; organised by Indian Group of IABSE, 2013, p. 10.

[5] F. Magalhães, A. Cunha (2011) Explaining Operational Modal Analysis with data from an arch bridge, Mechanical Systems and Signal Processing, Invited Tutorial Paper, Volume 25, Issue 5, pp. 1431–1450

[6] Emanuel Tomé (2019) Smart Structural Health Monitoring applied to Management and Conservation of Bridges, PhD Thesis, University of Porto

[7] A. A. Fonseca & F. M. Mato (2005) Infant Henrique Bridge over the River Douro, Porto, Structural Engineering International, 15.

www.ingramcontent.com/pod-product-compliance
Lightning Source LLC
Chambersburg PA
CBHW041325210926
51847CB00003B/47
9783857482076